Benjamin Pfluger

Assessment of least-cost pathways for decarbonising Europe's power supply

A model-based long-term scenario analysis accounting for the characteristics of renewable energies

Assessment of least-cost pathways for decarbonising Europe's power supply

A model-based long-term scenario analysis accounting for the characteristics of renewable energies

by
Benjamin Pfluger

Dissertation, Karlsruher Institut für Technologie (KIT)
Fakultät für Wirtschaftswissenschaften
Tag der mündlichen Prüfung: 2. Mai 2013
Referenten: apl. Prof. Dr. Martin Wietschel, Prof. Dr. Orestis Terzidis

Impressum

 Scientific
Publishing

Karlsruher Institut für Technologie (KIT)
KIT Scientific Publishing
Straße am Forum 2
D-76131 Karlsruhe

KIT Scientific Publishing is a registered trademark of Karlsruhe
Institute of Technology. Reprint using the book cover is not allowed.

www.ksp.kit.edu

Print on Demand 2014

ISBN 978-3-7315-0133-6
DOI: 10.5445/KSP/1000037155

Für Mudda

Preface

"The use of traveling is to regulate imagination by reality, and instead of thinking how things may be, to see them as they are."

Samuel Johnson

The publication of this thesis marks the end of a long journey. I would like to express my deepest gratitude to the people without whose support and guidance I could not have reached my destination. I am greatly indebted to Professor Martin Wietschel, who helped me avoid numerous pitfalls along the way without forcing me to take a specific path. I am deeply grateful to my mentor, Dr. Frank Sensfuß, who allowed me to mess around with his model and constantly forced me to keep "Gerda's law" in mind. I would like to thank Dr. Mario Ragwitz, whose lack of time brought me to the ISI and whose excess of inspiration kept me there. I would also like to thank my second reviewer, Professor Orestis Terzidis, for his thoughtful input.

This dissertation is the result of my work at the Fraunhofer Institute for Systems and Innovation Research ISI. The ideas and methods in it were generated in several projects, most notably the RESPONSES project, funded under the EU's Seventh Framework Programme. Without my colleagues at the Fraunhofer ISI, the occasionally necessary sprints (that surprisingly often turned into long-distance running) would not have been possible. It was their constant support, their ideas and the institute's atmosphere that kept me going. I would like to mention the following as representatives: Johannes Leisentritt for the nightshifts at the pony farm, Fabio Genoese and David Dallinger for bug-hunting, Jan Steinbach for being an exceptional office-mate and Helen Colyer for proofreading. Furthermore, I would like to thank Eiichiro Oda and the PWV for distraction.

Finally, I cannot even begin to thank Maren, for I lack the words to express my gratitude.

Like any journey, this one taught the traveller a lot about himself. For example, that metaphorical journeys are no substitute for actual ones and that, as Paul Theroux wrote, "travel is glamorous only in retrospect."

Contents

Contents

Abbreviations

ACER	Agency for the Cooperation of Energy Regulators
ECF	European Climate Foundation
EUA	EU Allowance Units
PV	Photovoltaics
RES	Renewable energy source
TGC	Tradable green certificates
AA-CAES	Adiabatic compressed air energy storage
ACE	Agent-based computational economics
BCE	Base case exchange
CCS	Carbon capture and storage
CGE	Computable general equilibrium
CHP	Combined heat and power
CO_2	Carbon dioxide
CPPM	Conventional Power Plant Manager
CPU	Central processing units
CSF	Conjectured supply function
CSP	Concentrated solar power
CW	Calendar week
DC	Direct current
DG	Directorate-General
DR	Demand response
DSM	Demand side management
EFF	Strengthened Efficiency
EGS	Enhanced geothermal systems
ENTSO-E	European Network of Transmission System Operators for Electricity
EPR	European Pressurized Reactor
ETS	Emissions Trading System
EU	European Union
EU-27+2	EU-27, Norway and Switzerland
FIP	Feed-in premiums

Contents

FIT	Feed-in tariffs
FLH	full load hours
GEP	Generation expansion planning
GHG	Greenhouse gas
GIS	Geographic information system
GP	Greenpeace
GRID	Hampered Grid
GUI	Graphical user interface
GW	Gigawatt
GWh	Gigawatt hour
IAM	Integrated Assessment Model
IEA	International Energy Agency
IPCCC	Intergovernmental Panel on Climate Change
kW	Kilowatt
LCOE	Levelised cost of electricity
LP	Linear programme
LPM	Linear Problem Manager
MA	Multi-agent
MIP	Mixed-integer problem
MOLP	Multi-objective Linear Programming
MS	(EU) Member States
MW	Megawatt
MWh	Megawatt hour
NO	Norway
NoCCS	No CCS
NREAP	National Renewable Energy Action Plan
NTC	Net transfer capacity
O&M	Operation and maintenance
OPT	Optimistic Decarbonisation
OTC	Over-the-counter
PBL	Netherlands Environmental Assessment Agency
PDTF	Power Transfer Distribution Factor
PHES	Pumped hydro-electric storage
PSM	Pumped Storage Manager
RE	Renewable energy
RES	Renewable energy sources
RES-E	Electricity generated from renewable energies sources

SD	System Dynamics
SFE	Supply function equilibrium
SQL	Structured Query Language
TGM	Transmission Grid Manager
TRM	Transmission Reliability Margin
TSO	Transmission system operators
TTC	Total Transfer Capacity
TWh	Terawatt hour
UCM	Unit Commitment Models
UK	United Kingdom
WEPP	World Electric Power Plant (Database)

List of Figures

List of Tables

1. Introduction

1.1. Background

In recent decades, the European electricity sector has fundamentally changed, and the restructuring process has been driven mainly by a politically forced market liberalisation. Although this process is still ongoing, fighting climate change has emerged as another and possibly even more challenging driver of change. Convincing evidence exists that global warming is anthropogenic and that its mitigation would be beneficial (IPCC, 2007a,b, 2008). Since the climate summits in Copenhagen and Cancun, the goal to keep the mean global temperature increase below 2° C until the end of the century is recognised by the international community (United Nations, 2009, 2010).

To quantify its contribution to reaching the 2° C target, the EU has set the target to reduce its greenhouse gas (GHG) emissions by 80 % until 2050 compared to 1990 levels (European Commission, 2011a). This goal, if it is to be seriously pursued, requires extensive changes in the electricity system: since the sector accounts for approximately one third of the EU's carbon dioxide (CO_2) emissions, its decarbonisation is considered mandatory in order to achieve the 2° C target.

The necessary or optimal decrease in emissions from the power sector essentially requires a trade-off of costs and benefits between this and other sectors. Previous research has concluded that the power sector offers a wide range of mitigation options at costs below many of those in other sectors. The European Commissioner for Energy, Günther Oettinger, summarised the decarbonisation target of the EU power sector as follows. "[...] if we have to make an overall reduction of 80 % that means that the energy sector has to bring its emissions to almost zero" (Oettinger, 2011). When postulating that this is correct[1] the questions remains how the power sector should be decarbonised.

---

[1] The actual reduction level is (of course) very relevant and is discussed in this thesis as well.

1.2. Problem definition

The discussion on anthropogenic climate change also initiated the search for feasible ways to mitigate it. Over recent years, many studies researched mitigation strategies for the energy sector, often developing different *2° C scenarios*, i.e. developments assessed to allow the 2° C target to be achieved. In these scenarios the reduction of the specific emissions from the generation of electricity are, at least in the long run, almost entirely achieved by replacing emission-intensive power from fossil fuels with:

- electricity generated from renewable energies sources (RES-E),

- nuclear power and

- power from plants equipped with carbon capture and storage (CCS) technology.

Decarbonisation scenarios are developed by institutes and researchers using different model approaches, implying that the results deviate in many key aspects. However, agreement exists that the share of RES-E must increase substantially. A comparison of scenarios for the decarbonisation of the European power sector sector by Fischedick et al. (2012) concludes that "all scenario studies [taken into account] whose scenarios run until 2050 indicate that the continent's electricity demand could be largely (at least by 80 %) or even entirely be met by a mixture of renewable energy sources by the middle of the century." In all scenarios a large proportion of the additional RES-E comes from wind and solar energy, i.e. fluctuating sources.

The diffusion of RES-E has a significant impact on other infrastructures; it is often seen as a central driver for the need to expand the electricity grid (ENTSO-E, 2012c) and spurs the discussion on the necessity of new electricity storage facilities. The utilisation of dispatchable power plants depends increasingly on the feed-in of fluctuating renewable energy source (RES). Capturing the characteristics of renewable energies is thus a central part of modelling the future power sector; an overly simplified representation of RES-E in the applied models could lead to unjustified or even wrong conclusions.

Nevertheless, strong simplifications are particularly necessary in models performing capacity expansion planning for power plants; keeping the size of models covering a large, diverse area (such as Europe) over a long time horizon manageable necessitates cutting back on temporal or spatial resolution and coverage.[2] Existing

---

[2]In this context *temporal coverage* refers to the number of time steps or system states a model takes into account. It ranges from a limited number of days assumed to be typical ("type day

modelling approaches are usually tailored to conventional electricity systems based on dispatchable power plants, in which both generation and demand follow relatively repetitive patterns. The appropriateness of such approaches seems disputable for a future power sector with high proportions of RES-E. On the one hand, capturing the characteristics of fluctuating RES requires a high temporal resolution and coverage to depict certain challenges, such as long wind calms. On the other hand, it seems necessary to model the European power system in its entirety, because the ability to balance fluctuations between weather regions is likely to be essential for systems based on RES.

1.3. Objective and approach

This thesis analyses technological least-cost pathways for deep emission reductions in the power sector of the EU Member States (MS), Norway and Switzerland. It seeks a better understanding of the role renewable energies play in the transformation process up to 2050. In particular, it focuses on the power supply side, taking into account electricity generation from both conventional and renewable sources, electricity transport and storage facilities. Therefore, it is a secondary objective of this thesis is to develop a model framework that allows for a realistic representation of fluctuating renewable energies and the resulting impacts on other system components. The model should be able to endogenously calculate capacity expansions for these infrastructures. Investment and unit commitment decisions should be based either on cost-efficiency or on profitability under a given market design.

The work starts with a summary of the developments of the European power sector in recent decades in chapter 2 . The chapter's objective is to provide an overview of the implications that the regulations and trends will have for future developments and for approaches to model these.

Chapter 3 discusses electricity sector modelling. It presents an overview of existing modelling approaches and their respective strengths and weaknesses. After defining the capabilities necessary for the task at hand, an assessment of the applicability of available models is given.

The analysis concludes that developing a new model is the most reliable approach, as it can be specifically designed for optimising capacity expansions for large regions while allowing a high temporal resolution and coverage. Chapter 4 introduces the components of the new model, PowerACE-Europe, and the formulation of the linear programme it defines. The chapter concludes with a critical reflection on the model.

---

approach") to covering all hours of the year.

In chapter 5 the input data for the four scenarios that are calculated with the new model is defined. The scenarios are designed to reflect current urging questions regarding Europe's power sector decarbonisation strategy.

Chapter 6 presents the scenario results. Since there are many indicators to be discussed, the initial discussion focuses on one of the scenarios which is evaluated in significant detail. The functioning and impacts of the particularities of the model are analysed. The remaining scenarios are subsequently discussed, focusing on deviations from the first scenario and tracing the differences back to their causes. Furthermore, a sensitivity analysis is performed and key results are measured against those of comparable model-based decarbonisation studies.

Chapter 7 provides a summary of the thesis, its central conclusions and an outlook.

2. Basic characteristics and trends of the European power sector

When generating and analysing possible techno-economic trajectories of the European electricity system, it is important to acknowledge that the system and the way it is organised and regulated has changed significantly over the last decades. In this chapter, these changes and their drivers will be briefly summarised. The objective is not to give a comprehensive description as this is beyond the scope of this work. Instead, the chapter provides a summary of the trends in the power sector and their implications for attempts to model the electricity system. The chapter focuses on *exogenous intervention* which, to a large extent, is the result of policy-making. Understanding exogenous influence is important for modelling exercises: Whenever a system is modelled, it is necessary to comprehend the "forces" working on it from the outside as well as their implications for the endogenous processes within the system. The analysis focuses mainly on developments that take place on a supranational, i.e. European level. Each of the aspects touched upon in this chapter is discussed in greater detail in other publications, to which the reader is referred in the respective section.

2.1. The European Union and its energy policy

The electricity sector has some features which distinguish the trade with power from other commerces. Usually, the activities necessary for supplying electricity to final customers are distinguished as follows:

- generation,
- transmission,
- distribution[1] and
- retail sales.

---

[1] In this definition, *transmission* usually encompasses the operation of the high voltage grid, whereas *distribution* refers to the operation of the lower voltage grid levels. Grid structures with a voltage level of 110 kV used to be seen as part of the transmission grid, but due to increased load densities have since become part of the distribution grid especially in urban areas (Heuck et al., 2007, p. 82).

The grid related activities are generally regarded as natural monopolies, since providing more than one infrastructure for the same service would not be cost-efficient. Furthermore, electricity is a perfectly homogeneous good that can be produced in a larger number of different technologies. Electricity is also more difficult to store than most goods, because bridging the temporal gaps between supply and demand is only possible with expensive storage facilities. Any significant mismatch in supply and demand results in outages, which can spread over large regions and cause immense costs, surpassing the market value of the unsupplied units of electricity by several orders of magnitude. Additionally, in most countries of the EU providing electricity is a major position in the country's GHG emission balance.

All of the four power service activities introduced above have been subject to significant changes in the last two decades. Although some of these changes can be described as "endogenous" changes[2], e.g. technological advances, the restructuring process was and continues to be actively driven to a large extent by political and regulatory influences. A large proportion of the intervention is directly or indirectly rooted in the legislation of the European Union and the directives of the European Commission.

The European countries have a long history of cooperating for coordinated intervention into the energy sector: Two of the three treaties defining the European Communities concern cooperation in the energy sector: the European Coal and Steel Community, established by the treaty of Paris in 1951, as well as the European Atomic Energy Community (Euratom), founded in 1957. However, it is also important to recognise that EU organs have only limited and indirect possibilities for intervening in the energy policies of its Member States. As the energy sector affects very sensitive sectors 'close to the state', the MS have always been reluctant to relinquish control over it (Eberlein and Grande, 2005).

The Lisbon Treaty (European Union, 2007) marks a new chapter of European energy policy: For the first time, it contains an "Energy" section; previously, there was no explicit definition of the EU's energy policy, e.g. under which conditions and how EU organs can set rules that are binding for MS. The EU was not able to act directly, as it can only act where it has been explicitly granted the competence to do so by the Member States. Intervention in the energy sector prior to the Lisbon Treaty was based solely on the EU's primary law and general competences, for example regarding climate action and competition law. The Lisbon treaty sought to make a clearer distinction of how the executive and legislative competences are

---

[2]It is difficult to clearly distinguish endogenous and exogenous effects for the electricity sector, because it is enmeshed with other sectors and strongly regulated.

distributed between MS and EU organs (Pielow and Lewendel, 2011). The treaty states that, regarding energy, it is the objective of the EU to:

> "(a) ensure the functioning of the energy market;
>
> (b) ensure security of energy supply in the Union;
>
> (c) promote energy efficiency and energy saving and the development of new and renewable forms of energy; and
>
> (d) promote the interconnection of energy networks." (European Union, 2007)

Organs of the EU can now intervene in the energy sector in a more clearly defined way. It could be assumed that the trend towards a more unified European energy strategy will continue in the future, rather than having different, often conflicting strategies of the MS. This is important when generating long-term strategy options and scenarios for the power sector, because these might require a strong willingness to cooperate.

The policies implemented so far have already demonstrated the willingness of policy makers to intervene in the energy sector in general and the power sector in particular. A major change initiated by the European Union applying secondary law is the liberalisation of the power sector.

2.2. Liberalised electricity markets

Of all the changes in the European electricity system over the last three decades, the still ongoing liberalisation probably has the most extensive structural impacts. The central motivation for liberalising electricity markets is to promote cost-efficiency through competition (Sioshansi, 2006). However, other aspects of a more strategic or political nature may also play a role. Pollitt (2009) defines four essential steps of an electricity market reform:

1. The privatisation of publicly owned electricity assets,

2. opening of the market to competition,

3. vertical unbundling of transmission and distribution from generation and retailing and

4. the introduction of an independent regulator.

In Europe, the debate concerning the necessity for electricity market reform gained momentum in the early 1990s. At that time, the electricity system, which could not yet be called a market, was organised in the form of national monopolies. The first real action of the EU took place in the *Internal Market in Electricity Directive* (European Commission, 1996a). The Directive stated that the several elements of the electricity system should be opened to the market, either immediately or gradually over time. A central measure of the Directive is *unbundling*, i.e. companies in charge of the transmission system have to be separated (at least in the management structures) from generation and retail. Furthermore, the MS were obliged to deregulate access to the grid infrastructure, opening it to non-incumbent companies. In some cases, the formerly state-owned monopolies had to be broken up or forced to sell parts of their power plant portfolio in order to comply with competition law. The retail market was deregulated and opened to competition. In most countries, this happened gradually.

The transformation into an internal market for electricity is far too complex to be covered here in detail. The reader is referred to a number of studies on the subject, e.g. Jamasb and Pollitt (2005) and Pollitt (2009), both of which analyse the progress of integrating the European electricity markets, the first chapter of Ferreira Dias (2011) on the same topic, or Schiavone (2010), who analyses the impacts of liberalisation and integration from the perspective of the MS. Furthermore, a large number of publications by the EU monitor the progress of liberalisation; noteworthy and influential reports include the annual benchmarking reports on the opening of the electricity markets [3] and the Energy Sector Inquiry of the Directorate-General (DG) Competition (European Commission, 2007)[4]. To sum up, the results of liberalisation have been disappointing, especially in the early years, when regulations even had distorting effects on the market (Jones and Webster, 2006). After initially promising results, end-consumer electricity prices did not show the desired decrease, and even increased in many cases (European Commission, 2007; ter Keurst, 2011). The amount of electricity traded between the MS increased only slowly, with little progress even in recent years.

The total electricity exported and imported as a share of the total electricity generation (Fig. 2.1) can be used as one indicator for market integration. The electricity traded between the MS (and Norway) increased only insignificantly over the last years, with the exception of 2011. Whether or not the increase in 2011 was merely the result of temporary shortage in generation capacity or marks the beginning of a trend will have to be analysed in the future. Along with the lack of trade

---

[3]The reports are available at http://ec.europa.eu/energy/gas_electricity/legislation/benchmarking_reports_en.htm.

[4]This report can be seen as one of the central triggers for stricter competition rules in the Third Energy Package.

comes a slow convergence of market prices. The observation that the cross-border flows have not increased significantly while there are considerable price differences between the countries' electricity markets has to be seen as a sign for lacking market integration. EU policy currently tackles this issue successively through *market coupling*[5] of regions. The EU focuses on regional, sometimes overlapping sub-markets as an intermediate step towards a single market. In recent years, at least for the Central Western Europe region as well as the Nordic markets, the increased connectivity and market liquidity has led to a greater degree of convergence (Huisman and Kiliç, in press).

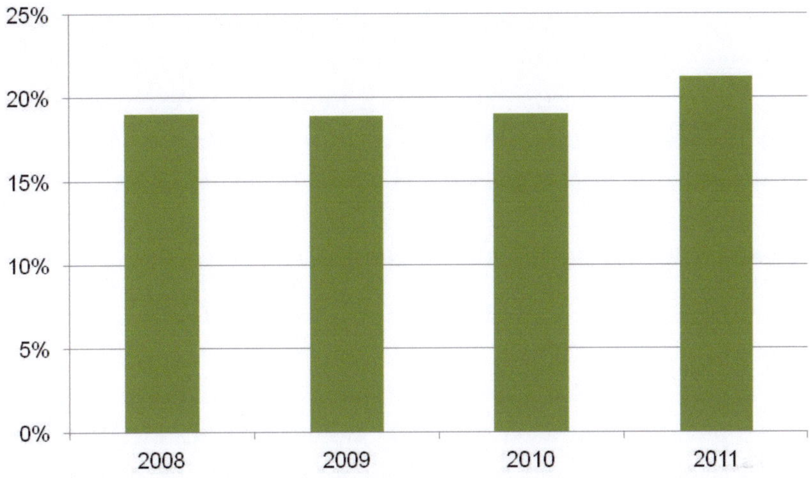

Figure 2.1.: Total electricity exports and imports as shares in total electricity generation in the EU-27 and Norway.

The most recent large set of regulations is the Third Energy Package, which essentially consists of five new legal acts, three of which directly address the electricity sector (see: European Commission (2009d,e,f)). The Third Energy Package represents a significant change, as, for the first time, energy and environmental aspects are treated together and addressed by the same measures. It is also much stricter and more explicit in its prescription of unbundling. A significant step towards creating a single market could be the creation of the *Agency for the Cooperation of Energy Regulators* (ACER). ACER's objectives are to ensure a cooperation between the national regulators, especially on regulations with cross-border impacts, and to

---

[5]In this context, market coupling refers to the implicit auctioning of interconnector capacities between two countries with different power exchanges.

advise the European organs. At present, ACER has legal authority only in technically necessary regulations, but the institution could conceivably develop into a "European regulator" in the mid-term future (Pront-van Bommel, 2011).

ACER can be seen as complementary to the *European Network of Transmission System Operators for Electricity* (ENTSO-E), which was founded in 2008, uniting the six regional associations of TSOs. ENTSO-E is not only important for being a platform for the Transmission System Operators (TSO) to define rules and harmonise grid extension plans, but also for transparency reasons: The organisation publishes key market results on its websites, e.g. physical and commercial flows of electricity.

It can be concluded that after a rather slow start, a trend towards market integration is emerging, not only for the MS, but also extending to Norway and Switzerland (Balaguer, 2011). Liberalization has led to a European market, which cannot be considered homogeneous, but which doe show signs of convergence. Certain market elements are similar in most countries, and will be briefly introduced below.

2.2.1. Trade on liberalised electricity markets

In the liberalised electricity market, electricity is traded either on power exchanges or over-the-counter (OTC). Power exchanges have become large-scale trading platforms and are key elements of the electricity market. On power exchanges, highly standardised products are traded. Leonardo Meeus (2011) distinguishes two types of power exchange: *Merchant exchanges* are set up either by utilities, financial companies or TSOs in order to make profits from their services. Profits are made through fees for various services, e.g. sold or bought volumes or sales of market data. The second type, *cost-of-service regulated exchanges*, are set up either by public institutions or by TSOs. The major difference is that these exchanges are either non-profit or their profit is regulated. Leonardo Meeus (2011) also points out that the regulated exchanges often carry out additional tasks, such as allocating capacity payments or performing congestion management. All exchanges allow the trading of electricity for short-term physical delivery, usually day-ahead. The price emerging from trade on these *spot-markets* is often seen as the wholesale price of electricity.[6] Furthermore, derivatives are traded, such as year-ahead futures or options. The volumes traded on derivatives markets have already surpassed the volumes traded on spot markets and tend to be less volatile (Karan and Kazdağli,

---

[6]The British electricity market is different in this regard, because it does not have one central market with prices that act as a price signal; electricity is traded through OTC contracts and on several, rather illiquid exchanges.

2011). When trading on power exchanges, market participants remain anonymous. The derivatives market has developed rapidly, but has only recently become subject of supervision. However, the transparency requirements affect only a few large market participants (Pront-van Bommel, 2011).

2.3. The influence of renewable energies

Liberalisation can be seen as the central driver for many changes in the European power sector over the last decades. Especially in the last years, the increase in power generation from renewable energies has become another major source of change. Switching from fossil or nuclear fuels to renewable energy sources is associated with several benefits: sustainability, especially in the face of climate change, increased security of supply, since non-renewable fuels are, by definition, finite, decreasing import dependency and the creation of local jobs. When the European Union first acknowledged the topic in the Green Paper on "Energy for the future: Renewable sources of energy" (European Commission, 1996b) and the following White Paper (European Commission, 1997), the focus was clearly on the aspired effects on energy dependency.

The decision to increase the share of renewable energies was formally established in the first *Renewables Directive* (European Commission, 2001). This was amended by a new version in 2009 (European Commission, 2009a), which sets binding[7] targets for the overall share of RES in gross final energy consumption. In total, the EU aims to reach a 20% share of RES in gross final energy consumption, which is part of the so called *20/20/20 by 2020 package*[8]. The efforts for meeting the target are shared between the MS in a *flat-rate/GDP/capita* approach: Half of the required increase is a fixed percentage to be met by each MS, while the other half is distributed according to the Member State's wealth, measured by GDP per capita. The Renewables Directive of 2009 also requires the MS to publish their strategies for reaching their respective targets in the *National Renewable Energy Action Plans* (NREAPs).[9] Although no binding sectoral target exists for electricity, examining the NREAPs shows that meeting the defined goals would result in an EU-wide RES-E share of 34.0 %. To meet their targets, the MS may use mechanisms called '*flexibility measures*', i.e. statistical transfers between MS as well as joint projects or support

---

[7] The first Renewables Directive set the same targets, but only indicatively.

[8] Besides the goal regarding renewable energies, the EU has committed itself to a 20% reduction in GHG emissions compared to 1990 levels and a 20 % improvement in energy efficiency.

[9] The NREAPS are published by European Commission (2011e) and discussed, for example, in M. Ragwitz et al. (2011) or Beurskens et al. (2011).

schemes. A country can also import renewable electricity from non-EU states, as long as a physical import of electricity into the EU takes place.

2.3.1. Support schemes for electricity generation from renewable energies

How the respective national target is reached, i.e. the choice of the support scheme, is the responsibility of the MS. Due to the fact that renewable energy (RE) technologies, with the exception of large-scale hydropower, are only under special conditions able to compete with conventional power generation under current market conditions, their diffusion will only take place if they receive additional support.[10] Following the categorisation of Haas et al. (2004), RE support can focus either on quantities or prices, and target either the investment into the technologies or the generation from them. Over the few last years, three general types of support schemes have been predominant:

Feed-in tariffs (FIT) are a price-based mechanism focusing on generation. Electricity qualifying as RES-E is remunerated at a fixed tariff defined by a government institution. The tariff is usually technology-specific and, in most cases, varies with other parameters, such as conditions on the respective site (Klein et al., 2010). As FIT can be adjusted to the real electricity generation costs of the technologies, the support can be designed to be cost-efficient. The drawbacks of the approach are that RE power plants are separated from the market price signals and the resulting RES-E volumes and pace of diffusion can only be controlled indirectly.

Feed-in premiums (FIP) are very similar to FIT; they are also price-based mechanisms focusing on generation. However, the price paid per unit of supplied electricity is influenced to some extent by the (spot) market price for electricity. Typically, a premium is paid on top of the spot market price. The premium is often floored and capped to reduce the risk for the investor and the costs for society, respectively. FIP gives the operators of RE power plants incentives to react to price signals.

Quota obligations with certificate trading are quantity-based mechanisms also focusing on generation. Government institutions define a RES-E quota to be met by certain market actors, e.g. suppliers. The quota does not have to be met physically, but by presenting the appropriate number of tradable green

---

[10]However, the competition is heavily biased by the past and ongoing subsidies for conventional power generation and the lack of internalisation of external costs (Jacobsson and Lauber, 2006).

certificates (TGC) that are issued for the generation of a unit of RES-E, often one MWh. Ideally, the system implicitly determines the most cost-efficient solution for reaching a certain target. In reality, however, the system has been found to be neither effective nor efficient, despite a recent increase in key performance indicators (see: Steinhilber et al. (2011, p. 118)).

Besides these support scheme types, other or supplementary support measures are possible, e.g. tenders, tax incentives or indirect support by penalizing non-renewable fuels. Figure 2.2 shows main support scheme currently applied by the MS.

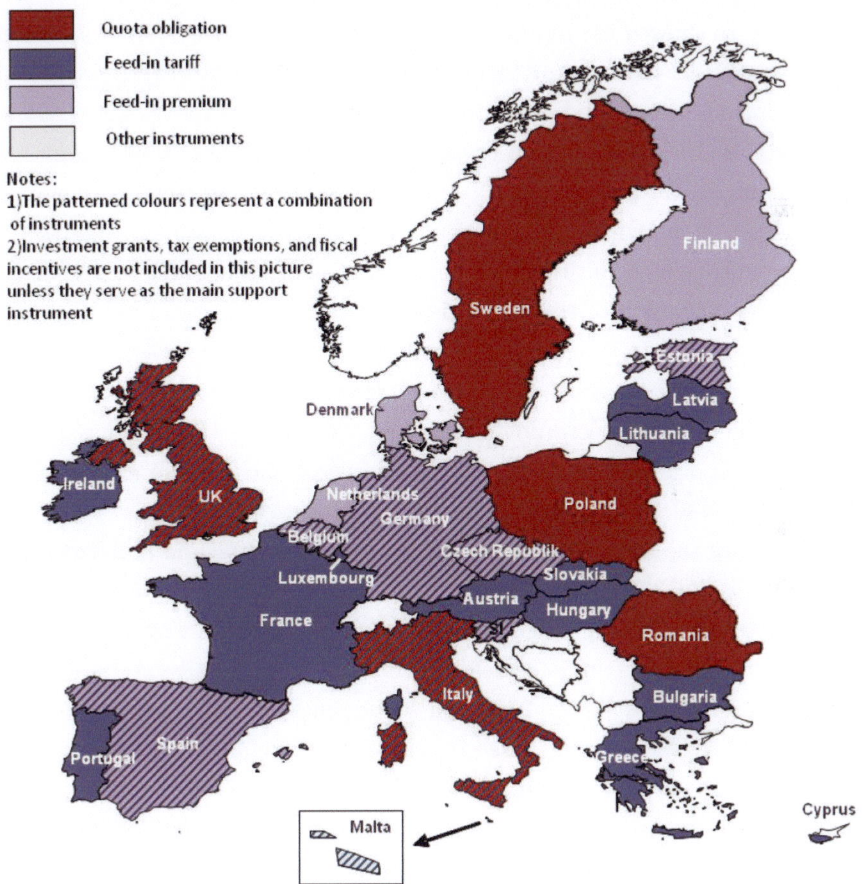

Figure 2.2.: Support schemes used in EU countries to promote RES-E. Source: Ragwitz et al. (2012).

2.3.2. Past developments and outlook

The recent, dynamic developments in the field of renewable energies in the European power sector are analysed by many publications, for example Jäger-Waldau et al. (2011). If the MS fulfil their respective targets stated in the NREAPs, electricity generation from RES will continue to increase strongly in the next years. Figure 2.3 shows the past as well as the planned developments documented in the NREAPs. It can be seen that only very limited growth is predicted for hydropower, both large- and small-scale, due to the relatively high costs of the former and the almost fully exploited potential of the latter. The highest growth is expected from wind power, which also showed the highest growth over the last years: Of the additional generation from RES between 2010 and 2020, 36 % and 23 % are expected to be from onshore and offshore wind power, respectively. A major expansion is envisaged for photovoltaics (PV), leading to a prognosticated generation of 83 TWh in 2050. In several countries, the diffusion of PV has been faster than expected. This might, however, be counteracted by the recent cuts in support in several countries, as several EU countries have put their RE support schemes on hold (Hamelinck et al., 2012) since the beginning of the debt crisis; the technological mix is increasingly discussed in the context of questions of affordability. However, strong growth is also planned for biomass, both solid and gaseous. However, the benefits and necessity of expanding biomass fuels are not undisputed, because of the interactions with the food sector and environment as well as the direct and indirect effects of land use change. This issue is summarised briefly by Tilman et al. (2009), while a more recent overview of the discussion taking place on the scientific and political level is given by (Kirkels, 2012).

The shares planned for the different energy sources illustrate why the challenges associated with the growth in renewable energies can be expected to increase: Over 70 % of the expected growth until 2020 comes from fluctuating RES. The generation from these technologies is characterised by a stochastic profile. While the respective plants can be curtailed if necessary, their generation is subject to the stochastic availability of the utilised energy source. This has implications for the rest of the electricity system that have to cope with increasing volatility. The technical and economic implications of this issue are not yet fully understood (Ambec and Crampes, 2012)[11].

---

[11]The authors of this paper refer to fluctuating RES as "intermittent" RES. This term seems inadequate, because the changes in produced electricity can be prognosticated, with an accuracy which increases with the number of generators (e.g. wind turbines), the area over which these are distributed and shorter forecasting horizons. The ramp rates that sudden calms create are less

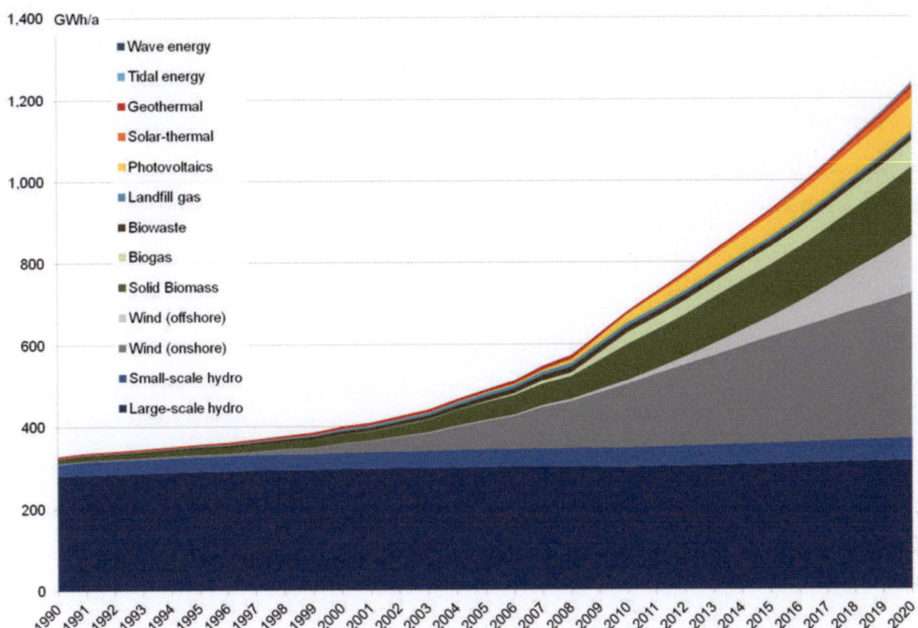

Figure 2.3.: Annual power generation from RE in the EU-27; past and future developments according to Eurostat and the NREAPs, respectively. Source: Own calculations based on Eurostat (2012) and European Commission (2011e).

The growth in renewable energies is already creating technical and economic challenges today. Some of the problems arise from the fact that most parts of the European electricity infrastructure system are designed for large, centralised, dispatchable power plants. Furthermore, current spot markets are *energy-only markets*[12] with still rather illiquid intraday markets, which complicates the market integration of RES-E.

So far, the countries are theoretically not bound by EU law or contracts to a further diffusion of the technologies after 2020. However, a continuation of the support for renewable energies and a further increase in power generation from RES-E seem likely. As will be shown in section 6.4, most researchers foresee a strong rise in the share of RE in electricity generation. If the developments in

---

steep, for example, than a technical failure of a nuclear power plant leading to a safety shutdown. The term "intermittent" often seems to be used in publications with a rather negative attitude towards RE.

[12]In energy-only markets, revenue is made solely through selling volumes of electricity. In capacity markets, the provision of capacity is remunerated. The are many combinations of both concepts in use. Whether or not capacity markets are needed in the face of increasing shares of fluctuating RES is currently being debated, see for example: Winkler (2012).

the field of renewable energies continued, it would be advisable, if not necessary to adapt many parts of the electricity sector. The increasing share of fluctuating generation needs to be considered when defining rules influencing investments into the rest of the power plant park or in the potentially necessary storage facilities, as well as the design of the transmission and distribution grids . Analysing these techno-economic interdependencies is a central objective of the work at hand.

2.4. Emission trading

Another policy intervention with high relevance for the energy and power sector is the *European Union Emissions Trading System* (EU ETS). The EU ETS was set in motion by the Kyoto process and defined in Directive 2003/87/EC (European Commission, 2003), which was amended in 2009 (European Commission, 2009b). The goal of the EU ETS (and emission trading in general) is to internalise the external costs of GHG emissions. The EU ETS is organised in a cap-and-trade form, with national caps negotiated between the Member States. *EU Allowance Units* (EUA), which represent the "right" to emit one ton of CO_2-equivalent, are allocated or auctioned by governmental authorities and can be traded on exchanges or OTC. Installations subject to the EU ETS have to hold sufficient EUAs to cover their emissions. The EU ETS is organised in phases, with the current Phase II covering the period from 2008-2012. While the power sector was subject to the regulation from the beginning of EU ETS trading in 2005, it has only been obliged to auction EUAs since the second Phase. A comprehensive overview of the origins of the EU ETS and an evaluation of the first phase is given by Ellerman and Buchner (2007).

The record of the EU ETS has to be seen as mixed. As can be seen in Figure 2.4, EUA prices reached an all-time high of approximately 33 EUR in spring 2006. However, the subsequent economic crisis caused the price to drop in 2008/2009. The average price this year was 8.31 EUR, up to and including September 2012. This means that the variable costs of a coal power plant with specific emissions of 650 kg/MWh$_{el}$ increase by 5.4 EUR/MWh$_{el}$ due to the EU ETS. Although this can have a impact on the utilisation of the power plant park, and interactions of the EU ETS with energy markets can be observed (Bredin and Muckley, 2011), price levels are not high enough to trigger investments in clean technologies (Rogge et al., 2011). The currently low prices could have several reasons, the most obvious one being the economic crisis. However, the allocation methods and the banking and borrowing options could also be at least partially responsible. In the third phase, which covers the period from 2013 to 2020, several changes will be introduced such as more stringent caps, full auctioning for the electricity sector[13] as well as the gradual inclusion of other sectors. However, many market participants expect the EUA price to surpass 20 EUR only towards the end of the third period (KfW/ZEW, 2012).

The developments in the EU ETS are difficult to assess for the medium- to long-term future. As the system originated from the Kyoto process, but is not directly linked to it (Ellerman and Buchner, 2007), it is possible that it could continue even without a post-Kyoto agreement, at least for a certain amount of time. Emission trading plays a central role in virtually all scenarios dealing with how to achieve the climate targets on a global level. In most publications on this topic, it is the dominant mitigation instrument. However, it only becomes effective at certain price levels. For this study the EU ETS, or a similar instrument pricing carbon, is assumed to continue to exist. Certain price levels will be assumed in the decarbonisation scenarios, to embed the European power sector in the assumed worldwide mitigation efforts.

2.5. Further exogenous influences on the power sector

Several other supranational policies influence the power sector, but can be only briefly outlined here. Most importantly, EU policy influences the demand for electricity in various ways. Noteworthy in this respect are especially the efforts to promote the efficient use of energy. This is reflected in the 20/20/20 by 2020 targets, that aim at a more rational use of energy. The energy efficiency and energy conservation policies of the EU encompass a range of directives and regulations. The main pillars are summarised and discussed, for example, in Wesselink et al. (2010). However, the results of the efforts to reduce the energy demand have fallen behind expectations. This is partly blamed on the *rebound effect*, i.e. that energy efficiency measures are often counteracted by direct or indirect effects (see for example: Herring (2006)), in some cases almost nullifying the measures. However, the size of the rebound effect is still discussed, for example, in a special issue of the Journal Energy Policy dedicated to the topic (see: France (2000)).

The continued growth in energy demand led to the formulation of a new *Energy Efficiency Plan 2011* (European Commission, 2011b), which proposes preferred ways how the different sectors can implement efficiency polices. Whether or not the

---

[13]In phase II, electricity generators have to purchase only a certain share of their EUAs by auction, while the rest was allocated free of charge.

Figure 2.4.: Developments of EUA price until September 2012. Source: Own calculation based on www.pointcarbon.com.

policies will be successful has relevant for this work as an exogenous influence on electricity demand.

After the disaster at the nuclear power plant "Fukushima Daiichi" in Japan in March 2011, several countries re-evaluated their policies regarding nuclear power. While some countries decided to react by making technical modifications to their nuclear reactors, others decided to (gradually) refrain from nuclear technologies completely. So-called *nuclear phase-out policies* were already in place in a number of countries, often as a reaction to the Chernobyl disaster in 1986. As of October 2012, Austria, Belgium, Denmark, Germany, Greece, Ireland, Italy, Portugal, Sweden and Switzerland either now have a phase-out policy or have never allowed power generation from nuclear energy. In Cyprus, Luxembourg and Norway, the construction of nuclear power plants is not restricted by law, but very unlikely for various reasons.

Another field influenced by supranational policies is the diffusion of CCS technologies. In this field, the CO_2 emitted from combustion processes is sequestrated, transported and stored underground in geological formations. Several different processes are being explored for the sequestration process. CCS gained a lot of its scientific attention with the latest reports of the Intergovernmental Panel on Climate Change (IPCC): From a global perspective it is difficult to generate long-term scenarios that are in line with the 2° C target without assuming the utilisation of CCS. The technology not only allows the utilisation of coal and natural gas with zero or at least low carbon emissions, it also allows *negative emissions*, i.e. the removal of CO_2 from the atmosphere (Rhodes and Keith, 2008). So far, the only scenarios developed without CCS are the " energy [r]evoltion" scenarios of Greenpeace (see, for example, the latest publication of this series, Teske, Muth, et al. (2012)), which achieve this through a very fast and radical transition to renewable energies in all sectors.

In the EU, this field is regulated by the *CCS Directive* (see: European Commission (2009c)), which focuses on administrative and regulatory clarification by the MS, but does not prescribe CCS targets or processes. Countries can exclude or forbid the utilisation of CCS on their territories. Currently, the technology's outlook is rather negative; many countries either banned CCS, or have difficulties implementing regulatory frameworks (von Hirschhausen et al., 2012).

2.6. Conclusions

This chapter introduced the basic characteristics of the European power sector, focusing on exogenous policy intervention on European level. Over the last two decades, the influence of the EU's regulatory framework on the energy policy of the MS has increased significantly. The Lisbon Treaty marks a new chapter in this development, as if officially recognises the energy market as a field in which the EU has a certain, albeit limited influence. Concrete technical aspects, e.g. the resulting supply mix, are still explicitly excluded from the competence of EU organs, and should be the result of market forces or national decisions. These changes and processes are modelled and analysed in this thesis. Therefore, the major endogenous "forces" should be represented adequately by the model. This chapter identified several trends in this regard. Among these, the increasing liberalisation and the various policies aimed at meeting the climate targets can be regarded as the most important.

3. Approaches to modelling the electricity sector

The purpose of the following chapter is to find a modelling approach suitable for the research question under consideration. It gives an overview of the different fields in which the electricity market and system models are applied together with the main modelling approaches. Furthermore, characteristics and features of models will be presented and evaluated for their applicability to the research question. The chapter aims to give an overview of existing work in the research field, defining the models suitable for the given task and assessing whether existing models can be applied.

3.1. Purposes of electricity market and system models

Over the past two decades, the efforts to analyse the electricity sector with computational models have significantly increased. The results of these efforts are visible in an increasing number of journal publications, and also in other literature dealing with certain aspects of the sector applying electricity market models. Political decisions regarding future pathways and framework of the electricity sector are increasingly substantiated by the results of computational models. For example, the German long-term energy strategy "Energiekonzept" is supported by the scenario calculations presented in (Schliesinger et al., 2010). On the European level, the Directorate-General for Energy recently presented different long-term decarbonisation strategies, for which quantitative impact assessments are presented in European Commission (2011d).

The increased interest in modelling the power sector is a result of several factors as their application is motivated by different objectives. When characterising the different modelling approaches, it is important to differentiate the fields of applications for electricity market models and the intentions behind their utilisation.

Firstly, the deep transformation of the electricity sector depicted in chapter 2 affects virtually all market actors. Many of the developments add complexity to the sector, which used to be organised centrally within individual countries. The most important source of the increased dynamism in recent years is the trend towards competition. Market participants have to define their short-term operations, e.g. unit commitment, and long-term strategies, e.g. investments, in the context of the actions of their competitors. Due to the unbundling of formerly vertically integrated companies and growing international competition, the number of actors and parameters to be taken into account increases. These developments are challenging for both market participants and governmental and regulatory institutions, striving to design an appropriate framework and supervise the market (Ventosa et al., 2005).

It is important to note that the intentions for modelling the electricity markets are different for these two groups. For market participants, electricity market models are primarily used as investment decision support instruments under a given framework, e.g. regulatory conditions. In turn, governmental bodies and non-governmental organisations apply them for market analysis and prognosis, for example testing alternate "global" developments, such as considered changes in the regulatory framework. Although these two types of applications are not completely distinctive, it is important to note the different objectives driving the modelling efforts.

Secondly, the electricity market is unique for several reasons which make it interesting from a scientific point of view. Electricity is a perfectly homogeneous good, although it can be produced in various ways differing in costs and technical limitations. Therefore, several market mechanisms can be keenly observed and without distortions: in the electricity market price differences cannot be explained by different qualities of the good but only by markets imperfections such as limited transfer capacity or the exertion of market power (Borenstein et al., 2000). Furthermore, trade and physical flow of electricity have to follow certain rules, since imbalances, even for very short periods of time, can lead to outages with severe consequences. Both the electrotechnical laws and the market rules are well-known or freely accessible, making the electricity system formally describable by applying these well-document rules. The degree to which these rules are included in the model can be varied according to the scope of the analysis and the computational resources at hand. For this reason, electricity market models are also of academic interest, for example in economics (ibid.). Although many models applied by market actors or governmental or non-governmental institutions are developed in, or are maintained by academic institutions, some models are used almost exclusively for academic purposes. The findings derived from these models can be valuable for understanding certain market mechanisms, but their results are often not directly transferable to real-world problems due to the models' simplifications.

Despite the fields of applications blending into each other, three broad and over-lapping fields of application can be identified for electricity market models:

1. decision support,

2. market analyses and forecast,

3. academic analyses.

Following the differentiation of Wietschel (1995), models designed for the first two fields can be classified as application-oriented, whereas models targeted at the last field can be seen as methodical. For the work at hand, the model is required to be rather application-oriented.

3.2. Model types

The following section will summarise attempts to characterise and categorise the most common approaches for modelling electricity markets. Generally, electricity market models have to be distinguished between *Unit Commitment Models* (UCM) and *Generation Expansion Planning* (GEP) models. UCM are used predominantly for short- and mid-term planning horizons. Several parameters such as installed power plant capacities and information about the electricity grid is exogenous to the model, whereas the dispatch of the power plants is calculated endogenously. In general, UCM tend to contain a higher level of detail in their market representation. If real-world capacities are used as input data, the results can be validated with market results and can be used to explain phenomena such as bidding strategies or market power.

In turn, GEP focus on the decision process for capacity extensions, e.g. power plants or electricity grids. Market mechanisms are often represented in a simpler way than in UCM models, especially for models covering a large region. This is due to the fact that the long technical lifetime of power plants of 30 or more years causes difficulties in including all aspects relevant for decision making, such as multiple markets or the competitors' behaviour over a long time horizon. Nevertheless, all GEP models deal with unit commitment to a certain degree, as the plants' utilisation has a large influence on investment decisions. Since the level of technical detail can be very high in GEP models, the problem is often simplified though reducing the number of analysed time steps to a limited number of exemplary situations, e.g. type days (see section 3.5.6).

Electricity market modelling approaches are often differentiated between *top-down* or *bottom-up* approaches (Enzensberger, 2003, p. 44). Top-down approaches consider technological aspects in an aggregated form, whereas bottom-up models emphasise representative individual technologies (Koch et al., 2001, p. 42). Top-down models deal with aggregated economic variables rather than detailed technical characteristics. The borders between bottom-up and top-down are ambiguous as models exist that apply a mixture of both approaches, which Krey (2006) refers to as 'hybrid models'. For example, Computable General Equilibrium models with a very detailed disaggregation of energy carriers and technologies could be described as hybrid models.

However, another classification of modelling approaches is possible, differentiating the models by their modelling focus: *Electricity market models* focus on all aspects of imperfect markets. Typical foci of the models are the strategies of market participants, market entrance barriers or the influence of market rules on strategies and prices. In contrast, *electricity system models* place greater emphasis on the developments in the whole electricity system. The representation of the market is less realistic and complex, allowing the analysis of larger system and longer time horizons. In common with the differentiation between bottom-up and top-down model, this classification is not strict; some models cannot be assigned unambiguously to one of the groups.

The main modelling approaches described in the following sections are depicted in figure 3.1.

3.3. Top-down modelling approaches

Typical examples of top-down modelling techniques are *Input-Output* models and *Computable General Equilibrium* (CGE) models (Koch et al., 2001, p. 47; Sensfuß, 2007, pp. 23-25). The basic theory of Input-Output models was developed by Wassily Leontief (Leontief, 1966), but has since been refined to explicitly include changes over time. The approach is also known as *macroeconomic modelling*, as the models apply the principles of macroeconomic accounting. Production and consumption are depicted through input and output tables, which can also take international trade into account. Input-Output models are mostly used for short-term to mid-term analyses, as the aggregation level renders the reproduction of technological change difficult (Koch et al., 2001, p. 47). Nathani (2008) shows that these shortcomings can, to some degree, be overcome by integrating other modelling techniques.

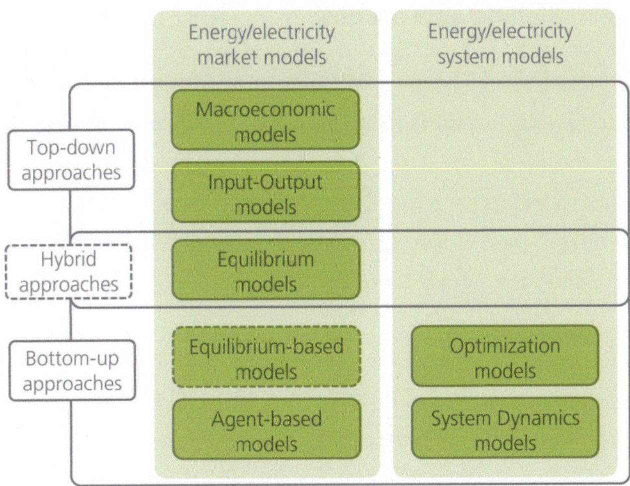

Figure 3.1.: Classification of well known approaches for modelling electricity markets. Own illustration based on Sensfuß (2007, p. 24) and Ventosa et al. (2005).

CGE models are based on neo-classical economic theory and depict the energy or electricity markets as being embedded in other national or international markets. The approach is related to Input-Output models, but puts a stronger emphasis on prices and their settlement. The name CGE is slightly misleading as the models do not necessarily rely on general equilibrium theory, as they can include market imperfections and externalities. Mitra-Kahn (2008) argues that CGE is not based on general equilibrium theory at all but should be seen as a subclass of macroeconomic modelling. In CGE modelling, supply and demand are matched by reaching balances in prices and quantities, with energy being regarded as a production factor. Depending on the aggregation level of the technical representation, some CGE models can also be classified as hybrid models (see section 3.4.1).

The main advantage of top-down-modelling is that linkages and interdependencies between sectors can be included and analysed. For example, the effect of increasing energy prices on the industrial creation of value or on employment can be depicted. The models can be used to include the effects on other sectors and occurring interdependencies.

As one of the objectives of the thesis is to depict the challenges arising from the integration of renewable energies on a technical level, it is a prerequisite to choose a bottom-up approach for the analysis. The following part of this chapter will focus only on bottom-up models.

3.4. Bottom-up modelling approaches

The existing approaches in the field of bottom-up modelling are often categorised into three major trends: *optimisation models, equilibrium models* and *simulation models* (Ventosa et al., 2005), which are rather distinctive.[1] In this classification simulation models encompass several modelling techniques that neither strive for a global optimum nor pass a strong emphasis on equilibria. The most prominent simulation approaches are *Agent-based computational economics* (ACE) and *System Dynamics* (SD).

The bottom-up approaches will be presented in the following, moving from concepts with a focus on markets, i.e. equilibrium and agent-based models, to electricity system models, i.e optimisation and system dynamics models. Each description starts with a brief introduction of the approach, followed by a summary of the features and advantages and is concluded by potential issues and known weaknesses.

3.4.1. Equilibrium models

Equilibrium models are characterised by their explicit consideration of market equilibria. The convergence of the the market is achieved by making use of several areas of economic theory, such a microeconomic theory, game theory, mixed complementary problems and mathematical programming with equilibrium constraints (Kagiannas et al., 2004, p. 417). The market participants, which, analogous to game theory, are often referred to as *players*, can be modelled with differing objectives. This allows for a often more realistic view of the market than that provided by models that assuming a perfect market without player's strategies. The models can be distinguished between *Cournot equilibrium* approaches and *supply function equilibrium* (SFE) approaches (Ventosa et al., 2005). In Cournot models, equilibriums are established by quantities, with the equations describing the market being algebraic. In SFE models equilibria are determined by both quantities and prices, thus allowing more complex strategies of the participants, which are described through differential equations. In most cases, convergence is reached through iterative algorithms. Both approaches seek for Nash equilibria, in which no actor can gain by changing his strategy unilaterally. Still, SFE approaches seem more suited to electricity markets, because the high price inelasticity of electricity consumers renders

---

[1] As it will be shown later on, some models exist that do not strictly match this classification and combine two approaches.

the pursuit of convergence through equilibrium quantities questionable (Day et al., 2002)[2].

Equilibrium models differ in the way that players react to changes in the market (Kagiannas et al., 2004) and how firms anticipate the reactions of their competitors to their own strategic actions. Day et al. (2002) give a comprehensive overview of possible formulations of these interactions in different market types. The players' strategies can incorporate a certain degree of technical detail, although many models have a long-term horizon, focusing on monthly or yearly equilibria. Equilibrium models are typically used for market power analyses (e.g. Kahn (1998)), and can cover both unit commitment and investments. For investment decisions, the sequence of the decision making process is of great importance,and is often used to analyse the possibilities of a company entering a market dominated by one or more incumbent companies. Iterative feedback loops can be used to connect the investment phase of the model with the unit commitment phase (Kagiannas et al., 2004). Furthermore, equilibrium models, especially SFE models can be used for *"obtaining reasonable medium-term prices"* (Ventosa et al., 2005), often surpassing the accuracy of optimisation models. This is particularly the case where the assumption of perfect competition is not justified, i.e. in markets where there are players with significant market power. In most cases the geographic scope of equilibrium ranges from small, hypothetical systems, to a small number of countries or regions.

In conclusion, equilibrium models can depict several particulars of real-world electricity markets. Nevertheless, incorporating these into models often comes with significant drawbacks: A major problem for all equilibrium models is convergence, as the conclusion of equilibria can only be guaranteed for rather simple problems (Day et al., 2002)[3]. Furthermore, the uniqueness of the equilibrium found cannot always be ensured, as several solutions often exist. Another issue pointed out by Ventosa et al. (2005) being especially relevant for SFE models is numerical tractability. Solving a large number of differential equations is very demanding in terms of computational resources. SFE problems describing real-world problems have so far only been solved under very restrictive assumptions. Furthermore, equilibrium models explicitly describe the players behaviour and interactions. Consequentially, many behavioural parameters which are not directly observable have to be defined

---

[2]Nguyen and Wong (2002) argue that this issue can be overcome by introducing an approach they refer to as *dynamic equilibrium* or *multiple equilibria* modelling. In this approach, demand is described through exponential-integral functions and the equations are solved using genetic algorithm techniques.

[3]The authors address these problems through conjectured supply function (CSF) models, which can be solved for larger problems. In CSF models, market participants conjecture on the reactions of their competitor to their own actions.

and configured (Day et al., 2002). For real world problems, one option is to iteratively determine these parameters until the model results match the observable data. Still, it is almost impossible to rule out the existence of other parameter settings with an equal or better fit. Even if plausible values for the parameters can be established, changes of these over time cannot be validated.

3.4.2. Agent-Based Computational Economics

Agent-based approaches seek to simulate the players' perspective on the electricity markets. The descriptions of the underlying structures are described on a microeconomic level, while macro-economic developments, e.g. taxation, can still be incorporated. The market actors are represented as agents analogous to the approach used in *Multi-Agent* (MA) programming. The term *agent* is not clearly defined in MA programming or ACE. Woolridge (2005) characterises agents as able to act autonomously, capable of interacting with other agents, reactive to their environment, while also being proactive. Leigh and Tesfatsion (2006, p. 835) give a more broad definition, in which "[e]ach agent is an encapsulated piece of software, that includes data together with behavioural methods that that act on on these data".[4] Agents have an explicit perception of their environment, according to which they decide upon strategies to follow. Their rationale can differ significantly as they can be generators, consumers, TSOs, market operators or other relevant entities. Although the approach is relatively new, being first discussed and applied in the early 1990s, it has quickly grown in popularity.

ACE tries to overcome three central limitations of equilibrium models and other bottom-up approaches. Firstly, ACE models do not dependent on a formulation that seeks for the emergence of equilibria. This allows the implementation of player strategies that do not aim to achieve a normative equilibrium (Sensfuß, 2007). In consequence, non-economic objectives and bounded rationality can be integrated relatively easily into decisions and actions of the agents. Secondly, in contrast to optimisation models, ACE models can deviate from perfect market assumptions in all aspects. Market power, imperfect or asymmetric information and entry barriers can be included and formally described. Thirdly, the approach is more dynamic than other approaches, as the agents can develop over time. This means that the agents learn through monitoring the impact of their strategies on their respective objectives. Adaptivity through the learning process is incorporated into many agent-based electricity market models. Agents are able to autonomously seek for

---

[4]One could argue that this definition is rather similar to objects in object-oriented programming.

their most successful strategy. Weidlich and Veit (2008) distinguish models by their applied learning algorithms:[5]

1. *Model-based adaptation algorithms* are tailored rather intuitively to the specific model and do not follow a standardised known formulation.

2. *Genetic algorithms* are heuristic methods inspired by biological evolution. The most successful strategies are determined by a "survival of the fittest" selection process, which can even include crossover and mutation processes to generate new strategies not defined in the initial model setup.

3. *Erev-Roth reinforcement learning* was suggested by Erev and Roth (1998) and is based on the learning processes of human individuals in repeated games. This approach is very common in agent-based electricity market models the bidding strategies of agents representing electricity generators. The agents experiment by bidding above the marginal generation costs through probabilistic functions.

4. *Q-Learning* calculates the expected utility of performing a certain action under certain circumstances. A typical example is a strategy in which the agent selects the action with the highest expected utility in a large proportion of actions, while randomly deviating in a small number of actions (ε-greedy-strategy).

Although the high flexibility integrated through adaptivity is seen as a valuable tool for overcoming static strategies applied in many other approaches, it comes with certain requirements for restricting the non-observable properties. As an example, Sun and L. Tesfatsion (2007) analyse a five-node transmission grid case with market rules similar to those ones of the electricity markets of the United States. By applying Erev-Roth-learning, the agents quickly adapt their strategies to bidding at three times their actual marginal costs. Although this result shows the possibilities of agents to exercise market power in this particular market, it fails to deliver realistic prices. This is a well-known phenomenon for all models applying game-theoretic elements: market actors with market power will exercise it to its full extent. In reality, the bidding behaviour of market participants is monitored by regulatory agencies such as competition regulators. In practice, market power can

---

[5]The algorithms presented in Weidlich and Veit (2008) furthermore include *Leaning Classifier Systems* and *supply function optimization*, which are not presented here. It is also interesting to note that the authors define adaptivity as a mandatory feature of agents in ACE, though there are models without endogenous learning that can be seen as agent-based.

be exercised only within certain borders. These limitations can either not be represented in the model or only be depicted on the basis of non-observable variables that have to be arbitrarily conjectured.

3.4.3. Optimisation models

The main characteristic of *optimisation models* is the existence of a single objective function (Ventosa et al., 2005), for which an optimum is determined under a set of restrictions. These typically cover both technical and economic limitations, which are called bounds. Optimisation models can either cover the energy sector as a whole, or focus on particular parts, such as the electricity sector. In both cases, the macroeconomic framework is modelled exogenously, thus disregarding potential intersectoral feedbacks (Koch et al., 2001).

Optimisation models were originally used as *single firm models* to analyse the electricity market from the view of one market actor. The objective is either the maximisation of profit or the minimisation of costs for delivering services. Prices can theoretically be determined endogenously, but are often assumed to be exogenous under the assumption that the single firm acts as price-taker. The problem is thereby reduced to a problem of bidding optimal quantities (see for example Gross and Finlay (1996)).

Additionally, single firm models also often deal with insecurity, for example of prices (e.g. Rajamaran et al., 2001). Further insecurities that can be taken into account are the behaviour of the competitors (e.g. Baíllo (2002)) or other stochastic variables, such as the inflow into hydropower systems (e.g. Unger (2002)). Several recent publications focus particularly on the insecurities in the investment decision for renewable energies: S.-E. Fleten et al. (2007) analyse the optimal point in time for investments into wind energy. Fuss, Szolgayová, et al. (2012), focusing on several low-carbon generation technologies, apply a firm-level optimisation process based on real options, in which the price for emission allowances is the major source of insecurity.

Depending on the degree of insecurity, optimisation problems can become very large. A common approach is to simplify the large equation matrix problem through Benders' decomposition, developed by Benders (1962). In Benders' algorithm, the large non-linear, often stochastic original problem, called the "Master problem", is divided into sub-problems, the solutions to which are used to update the Master problem and decrease the size of the solution space until a more profitable solution cannot be found (Kazempour and Conejo, 2012). An overview of stochastic modelling approaches in electricity market optimisation models is given in Krey (2006) and Möst and Keles (2010).

The second large group of optimisation models consider large electricity systems. In this case the objective of the optimisation is to minimise the costs occurring for meeting a typically exogenous demand, or the maximisation of welfare. In principle the same rules and restrictions apply as to single firm models. Due to the large size of the optimisation problem, linear optimisation is most often applied (Enzensberger, 2003). In the absence of differences between retail and wholesale prices the objective function can be seen as a centralised single-firm producer trying to minimise its cost (Nishimura et al., 1993). Most approaches assume an endogenous demand, as feedbacks to other parts of the economy are not included. However, calculation of market prices is more common, since the larger parts of the market are analysed. Optimizing electricity system models implicitly assume perfect competition (Nyober, 1997) and in most cases perfect information, though they are often myopic. The models can be applied to a large range of research questions, which leads to a variety of models with very different foci. An overview of well-known existing models is given by Krey (2006) and Enzensberger (2003).

Aside from models using a single objective function, *Multi-objective Linear Programming* (MOLP) has been used in several attempts to move away from a solely economic view on the electricity system. One of the first approaches in this context is explored in Clímaco and Almeida (1981). The optimisation methods used in MOLP define an objective that contains other goals besides a pure minimisation of costs. These might include, for example, environmental aspects or metrics for measuring security of supply. In a first step, non-dominated solutions are determined, thus building a Pareto front, which can be analysed or ordered, e.g. by Analytical Hierarchy Processes (Meza et al., 2007). MOLP is currently used predominantly for capacity planning from an academic or governmental perspective.

A central strength of optimisation models is the possibility of including a very high technical detail level for all components of the considered system (Sensfuß, 2007). For example, the technical behaviour of power plants in terms of ramp rates during start-up or shut-down processes, minimum uptime and downtime as well as partial load behaviour can be described formally in the form of a mixed-integer problem (MIP). Although ramp rates can also be included in linear problem formulations, minimum down- or uptimes cannot.

Another advantage of optimisation models is the transparency of the approach. Other modelling approaches can produce artifacts caused solely by certain model assumptions, for example by an assumed behaviour of market participants. It is often difficult to distinguish these artifacts from robust results, thus requiring rigid sensitivity analyses of the assumptions involved, which can consume a considerable amount of time. In turn, the highly simplified behaviour of market participants in optimisation models allows an easier interpretation of the results.

Nevertheless, the simplicity of optimisation models is also a point of critique. Firstly, optimisation models are *partial models*, viewing the energy or electricity sector detached from other parts of the economy. Furthermore, the approach has disadvantages in its representation of the market, as the assumption of perfect competition is a serious limitation. A high number of model-based studies research the effects of market power and other market inefficiencies, e.g. Hobbs et al. (2000), Wen and David (2001) or Bunn and Oliveira (2003). The research shows clearly that current energy markets are imperfect and offer significant incentives for market participants to deviate from a behaviour under perfect competition. Deviations can occur for example by market participants holding back capacity at certain times to generate scarcity, or through offers at prices above marginal costs, both of which cannot be depicted by optimisation models. As information cost, transaction cost and other market failures are not included, the approach tends to underestimate the cost related to systemic changes (Sensfuß, 2007). Another known weakness of optimisation models is the "Bang-bang" or "Penny-switching" effect, in which "small changes in input parameters might lead to considerable modifications in the output" (Held, 2011).

3.4.4. System dynamics

The SD approach was developed in 1959 by the electrical engineer Jay Forester (Forrester, 1958). The approach was initially used to understand the dynamics of stocks and flows in industrial processes and was only later found to be useful for modelling energy markets, with the most prominent example being "Limits to growth" (and subsequent studies) by Meadows et al. (1972). SD models are described through (non-linear) differential equations describing stocks and flows as *casual loops*. Casual loops describe interdependencies between monitored variables and are often described by response time into account. The feedbacks can be grouped into positive (amplifying) and negative (dampening) interrelations (Sterman, 2000).

SD can be seen as an answer to existing issues with equilibrium and optimisation approaches by focusing on developments in non-equilibrium states of the system. It makes the approach very interesting for the electricity sector, in which such dynamics often occur.[6] The majority of the publications applying SD in the electricity sector focus on capacity expansion. The "pull" for new generation facilities is generated by real or perceived scarcities in supply, which drive prices up and incentivise investments (Olsina et al., 2006).

In recent years, several SD approaches explicitly deal with investment processes in renewable energy technologies. An advantage of SD in this context is that the concept enables the depiction of the results of imperfect foresight, e.g. uncertainties regarding demand and fuel prices (ibid.) and uncertainties regarding meteorological conditions. This is performed for example by Hasani-Marzooni and Hosseini (2011), who simulate GEP and apply Monte Carlo techniques for modelling uncertainties in demand and wind speeds. The level of technological detail of the system description can be very high. The equations used in SD can include endogenous technological learning through learning curves and real options approaches (Kumbaroğlu et al., 2008).

Compared to optimisation models, SD allows for a better integration of market rules and the consideration of imperfect markets. Nevertheless, the complex interactions between the decision rules can lead to implausible results in long-term analyses with structural changes (Enzensberger, 2003).

3.4.5. Multi-approach models

Notably, some research applies models that cannot clearly be related to one of the approaches described above. This is the case for example for the model used in Centeno et al. (2007), which applies an equilibrium model that finds a solution by simplifying the equations to an equivalent optimisation problem. Other models apply two or more approaches to analyse a market from various angles. A notable concept in this regard is for example Jin and Ryan (2011), who apply a bi-level equilibrium model that can either search for a global optimum using three different optimisation algorithms, or allow strategic behaviour of market participants. Even without an explicit coupling of the models, analysing a given case with multiple models follows similar objectives. This is demonstrated, for example, by Lund et al. (2007), who sequentially use an Input-Output model and an optimisation model for analysing the electricity system of a Croatian Island.

Multi-approach models are interesting from a scientific point of view, as they allow a better validation of, and differentiation between solid model results and artifacts

---

[6]The developments after the retirement of a large-scale power plant, generating a certain shortage of supply and thus incentive for (temporally delayed) investments are an example of a non-equilibrium state.

of the respective modelling approach. They can also be relevant for policy-advice, enabling a comparison between an idealised, optimal market with the lowest possible costs and an imperfect market. This comparison can provide valuable information on how to counteract the distortions through regulations or political intervention.

3.5. Model characterisation and capabilities

The selection or creation of an appropriate model strongly depends on the research question. The selection does not only concern the modelling approach, but also has to take into account requirements regarding the capabilities of the models. The following section will present the different capabilities of electricity market models. An assessment is made on whether the capability is required to answer for answering the research question of this work. An overview of the capabilities and the results of the assessment is given in table 3.1. In the table, the necessity for integrating a certain aspect is categorised as follows:

1. - *Not applicable*: The capability or characteristic is regarded to be unsuited for the given task.

2. *+/- Optional*: The capability or characteristic is an option for the given task, but other approaches are also conceivable.

3. *+ Subsidiary*: The capability or characteristic is expected to bring additional insights, but is not seen as mandatory. Whether the capability should be integrated into the model depends on a weighting of potential benefits and the impact on computational properties such as calculation time and memory requirements.

4. *++ Necessary*: The capability or characteristic is expected to be essential for the given task. Integrating the capability is seen as a *condicio sine qua non*.

Decision variables in this context refers to those endogenously determined by the model. As explained in section 3.2, models can be distinguished between Unit Commitment Models (UCM) and Generation (capacity) Expansion Planning (GEP) models. GEP models are used mainly for three types of analyses: Firstly, to find an optimum strategy for specific investment decisions. Secondly, for modelling investments decisions in short- to mid-term analyses. In this case, the focus is often on market power or the influence of uncertainty, as is the case, for example, in Centeno et al. (2007). Thirdly, GEP models are used for long-term scenarios, for

Table 3.1.: Characteristics and capabilities of electricity market models and their applicability to the task in this thesis

Characteristic	Examples		Necessary for the research question
Modelling approach	Marco-economic	-	Not applicable
	Equilibrium	+/-	Optional
	Optimisation	+/-	Optional
	Simulation	+/-	Optional
Decision variables	Unit commitment	++	Necessary
	Generation extension	++	Necessary
Commodity markets covered	Electricity	++	Necessary
	Gas	+	Subsidiary
	Coal	+	Subsidiary
	Heating/cooling	+	Subsidiary
Electricity markets covered	Spot market	++	Necessary
	Future markets	+	Subsidiary
	Reserve markets	+	Subsidiary
	Intraday markets	+	Subsidiary
	Retail market	+	Subsidiary
Geographic area	Single project	-	Not applicable
	Local/community	-	Not applicable
	State/national	-	Not applicable
	International	++	Subsidiary
Electricity grid	Single node/no grid	-	Not applicable
	Transshipment model	++	Necessary
	DC flow	+	Subsidiary
Time step	Seconds	+	Subsidiary
	Quarter of an hour	+	Subsidiary
	Hour	++	Necessary
	Month/years	-	Not applicable
Time step coverage	Type days	-	Not applicable
	Type weeks	?	Uncertain
	Complete years	+	Necessary?
Players' market behaviour	Single generator	+/-	Optional
	Competitive strategies	+	Subsidiary
Uncertainty modelling	Completely deterministic	+/-	Optional
	Probabilistic demand	+	Subsidiary
	Probabilistic weather	+	Subsidiary
	Probabilistic power plant behaviour	+	Subsidiary

both scientific research (see for example: Olsina et al. (2006)) and for policy advice (see for example: Antoniou and Capros (1999), European Commission (2011d) or Lienert and Lochner (2012)). While some models only analyse investments into certain infrastructures, e.g. Torre et al. (2008), focusing on investments in transmission grids, others take several infrastructures into account, either simultaneously or consecutively. In this field, GEP models are useful as they allow the forecasting of changes of the electricity system and simulate the behaviour of the altered system.

Long-term GEP models usually incorporate a lower level in either technical detail or market representation than UCM. Modelling investment decisions of infrastructures with a long technical lifetime can, depending on the complexity of the decision process, be very demanding in terms of calculation time and required main memory. As computational resources and time available for calculations are limited, the trade-off between a detailed representation of the investment decision processes on the one hand and unit commitment on the other hand has to be considered carefully.

One of the fundamental assumptions of this thesis is that the challenges arising from fluctuating generation from RES are not sufficiently accommodated in the approaches currently applied in most models. For approaching this question the application of a GEP model is necessary; to define a realistic low-carbon emission electricity system, all infrastructure components have to be matched. Therefore, reliance on generation capacities published by other models or studies is deemed insufficient, as unit commitment and capacities have to be aligned. Due to the assumed complexity of a low-carbon system with a high proportion of fluctuating generation, modelling of unit commitment has to be performed by the model with a significant level of detail.

3.5.1. Commodity markets covered

Energy models can be distinguished by the energy markets covered, i.e. whether they just include a detailed representation of the electricity market or also simulate markets of other energy carriers. Multi-market models analyse the interdependencies between the markets for electricity, gas and coal and in some cases even the demand for heating or cooling. Most bottom-up electricity market models are partial models, modelling the electricity sector endogenously, while data input for other sectors or commodities included in the model is exogenous. An example of this is fuel prices: most models disregard price linkages between supply and demand. However, as Lienert and Lochner (2012) argue, argue, interdependencies exist and have been increasing over recent years especially due to the growing liberalisation of the gas markets.

Capturing the linkages between the electricity sector and the markets for fuels used for electricity generation requires other sectors using the same fuels, for example the industry and household sector, to be represented. Due to the high level of detail required for modelling the electricity sector and the focus of this work, it is concluded that this is beyond the scope of the study.

3.5.2. Electricity markets covered

As depicted in chapter 2, the liberalisation of the electricity market and the privatisation of infrastructure assets such as power plants, has led to both a higher integration and harmonisation between the European Member States and its neighbours, and also to the emergence of different markets. Electricity is traded as commodity both on different power exchanges and in different contract types. The most common market form are spot markets, which are operated in similar ways by almost every power exchange. Although most spot markets have gained liquidity in recent years, a certain share of contracts is still traded OTC, with the share varying from country to country. Besides the spot market, derivatives, most notably futures, are traded on separate platforms. The settlement prices of these have become influential indicators for trends and developments in the sector. Furthermore, in most countries reserve markets exists, although the regulations applied in these are rather heterogeneous. As explained in chapter 2,although intraday trading is assumed to become increasingly important in the future, currently the markets are illiquid compared to other markets and do not exist in all countries.

The current condition of the markets is reflected in electricity market models, as virtually all electricity market models cover the spot market(s) in one form or the other. Many models comprise the assumption that power is only traded on one or more spot markets, thus disregarding over-the-counter trade. One of the reasons for the strong focus on the spot markets lies in the high volumes of power traded in them and their importance as price signals for other markets, including OTC trade.

However, a challenge in this context is the long time horizon that needs to be covered by the model. The electricity markets described in chapter 2 are essentially still tailored to power generation from conventional or dispatchable sources[7]. It cannot be taken for granted that the the same or even a similar system will be

---

[7]The Nordic power markets already have a very high share of renewable energy. Still, due to the high dispatchability of storage hydropower, Nordic markets are organised in a similar as other markets in the rest of Europe.

applied when a large share of the generation relies on fluctuating energies sources with marginal generation costs of close to zero. Applying primarily energy-based pricing mechanisms, as it is the case today, could lead to several problems in the future, e.g. high uncertainty of investors on the revenues for peak power plants (Pöyry, 2011). As Bode and Groscurth (2011, p. 64) and Winkler (2012) point out, various market design options have been proposed to better account for the technical features of renewable energies. It is highly uncertain which of the options will be implemented and when this will be the case. For the modelling of long-term developments this means that both keeping the current market design, or arbitrarily choosing a different one for future years is highly speculative.

3.5.3. Geographic area

As already touched on above, electricity market models cover different geographic areas. Literature is rich in analyses dealing with either small, often hypothetical systems (consisting of one to 20 nodes) or a small number of countries. Less experience exists with large, real systems covering more than three countries in bottom-up models. Recent examples of pan-European modelling efforts are Teske, A. Zervos, et al. (2007), A. Zervos et al. (2010), ECF (2010), European Commission (2011d), Schaber et al. (2012) and Lienert and Lochner (2012). Besides the research carried out for and by the European Commission, which relies on the primarily equilibrium-based PRIMES model, all these analyses are carried out using optimisation models. Furthermore, following the differentiation of Wietschel (1995), all approaches could be characterised as application-oriented. [8] The reason for most models covering large areas over a long horizon being optimisation models applying rather simple assumptions regarding markets, lies in the high computational requirements of such tasks.

Another reason could be the inhomogeneity of the European electricity system in terms of electricity supply and market rules. To demonstrate this, it is valuable to compare approaches used for analysing the Scandinavian or Swiss electricity system (e.g. S.-E. Fleten et al. (2007)[9] or Unger (2002)) which are both systems with a high share of reservoir hydropower, with models analysing countries with a fossil-fueled power plant park. The models used for the Nordic power systems place emphasis on the often stochastically modelled inflow into hydropower plants and the players'

---

[8] The approaches used in these studies are not discussed in detail here, as a comparison is performed in chapter 6.

[9] An overview of models used for analysing Nordic electricity markets can be found in Vogstad (2004, pp. 17-23)

strategies induced by this insecurity. The applied modelling concepts differ significantly from those covering regions with a high proportion of nuclear power. For the latter, short- to mid-term prices can be derived for example by using periodic autoregressive models (Koopman et al., 2007), as real-world prices are fundamentally difficult to explain using marginal generation cost. Furthermore, technical regulations and market rules differ among countries and often change, thus requiring short refresh periods to keep the model updated. The comparison shows that most methodical model approaches are tailored to the system analysed. Therefore, the high heterogeneity of the European electricity systems renders approaches to capture it in one consistent model difficult and requires simplification and homogenisation

Nevertheless, in order to answer the research question, a model covering all Member States as well as Norway and Switzerland is necessary. Regional differences need to be included as far as possible. However, it is likely that the regional results within a pan-European model will not be comparable with results of models tailored for specific regions. In turn, the advantage of pan-European models is their ability to dynamically incorporate trade or physical exchange of electricity. Increasing proportions of fluctuating generation from RES especially highlight the need for to view the electricity system in its international context, as fluctuations in generation can be balanced between regions and countries. Finding ways to incorporate these inter-regional balancing mechanisms is a central motivation for this thesis.

3.5.4. Electricity grid

Several concepts exist for including grid aspects into the electricity market or system models, and their selection has a significant influence on the models' accuracy and computation time. It has to be noted that even rather detailed concepts are still a simplified depiction of the complex processes carried out in order to to keep the electricity system continuously stable. In electricity market models, there are essentially three different types of approach. These are briefly introduced here in ascending order of complexity:

1. *Single-node models* are the simplest approach; the endogenous calculation of grid aspects is essentially disregarded in the model. All generation and consumption of electricity is assumed to take place at one virtual point. The flow of electricity is not restricted by physical transmission constraints. Therefore, the concept is also referred to as "copper plate approach". Despite being strongly simplifying, the approach is very common for both academic research and for models analysing systems in which network constraints are assumed

to play only a minor role. This is the case for countries with a well-developed electricity grid, in which internal congestions seldom occur or do not greatly influence the market results. For example, the model developed by Sensfuß (2007) focuses on the German wholesale and reserve power markets. Although internal grid congestion can occur in Germany, the price of the wholesale market is not affected by it, rendering a potential inclusion of the electricity grid ineffectual.[10]

2. *Transshipment models* (also called transport models) take the location of supply and demand into account, they apply a simplified model for the dynamics of the electricity grid. In transshipment models, electricity behaves like a normal good and can be transported from one node to another. For each node, the sum of imports and production, less exports and consumption, has to be zero at all times. Therefore, transshipment models are in accordance with Kirchhoff's current law. The flow of energy is restricted by the transfer capacities of the power lines and losses are assumed to occur as a fixed share of transported power.

   Transshipment models do not explicitly consider voltage levels. In such an approach the voltage level of a certain power line influences its transfer capacity, but other aspects, such as voltage stability, are not modelled. Therefore, the approach does not take Kirchhoff's voltage law into account, i.e. that the sum of of all voltages around a closed network is zero. This can lead to the model suggesting system states not feasible in reality, leading to voltage instabilities or blackouts.

   Nevertheless, the approach is used for its simplicity and low demands in terms of computational resources and calculation time. To the best knowledge of the author, all long-term scenario studies covering Europe, e.g. Teske, A. Zervos, et al. (2007), ECF (2010) and Lienert and Lochner (2012)[11] apply transshipment models.

3. *Direct current (DC) power flow models* allow a more realistic representation of electricity grids and its flows on it, as they allow a differentiation between commercial and physical flows. The difference is illustrated in the analysis of

---

[10]The same model is also applied by Pfluger (2009) for an analysis of the Italian electricity market with explicit incorporation of grid constraints. The Italian market is set up in a way in which congestion can influence settlement prices, as the Italian market then splits into price zones.

[11]The model used in this study is not explicitly stated to be a transshipment model, but the description of the model suggests that this is the case.

flows in an electricity grid without international trade. In this case, electricity flows occur on trans-border lines without commercial trade taking place (Adamec et al., 2009). Very often, the flows can occur in opposite directions (Purchala, Haesen, et al., 2005). Electricity does not necessarily follow the path of the financial transactions, which are the centre of most electricity market models. The DC flow models are a simplification of alternating current (AC) flow models, in which the flows of both active and reactive power are simulated.

Several concepts of DC flow models exist and have been discussed and refined for decades. An early summary of existing concepts is, for example, given in Deckmann et al. (1980). A very common concept is that of *Power Transfer Distribution Factor* (PDTF) matrices. A PDTF matrix describes the physical flows that follow an economic flow on the basis of the impedances ("Extended factors for linear contingency analysis"). A PDTF is usually derived by analysing power flows in an AC model in several states and deriving conclusion about the correlations between physical and commercial flows. In contrast to AC models, the PDTF is linear, which allows their implementation in electricity market models. Purchala, L. Meeus, et al. (2005) define a set of criteria applicable to a network that allows a DC flow simplification for techno-economic analyses. The same authors then conclude that these criteria are met for the European electricity network for which they develop a DC flow model (Purchala, Haesen, et al., 2005).

For the work at hand, using a single node model for the whole European electricity system is obviously not applicable, whereas the utilisation of a transshipment or DC flow model both seem fitting. The advantage of an DC flow approach is that it delivers more realistic results as the power flows are calculated in accordance with Kirchhoff's law. The disadvantage is that for simplifying a problem through a PDTF or similar approach, extensive knowledge of the infrastructure is required. This means that the existing transmission grid and its nodes have to be incorporated in the model and system load profiles and power plants have to be assigned to the respective nodes. Due to the long-term horizon of the research question, the most likely significant changes would also have to be depicted. Even with excellent computational resources it currently does not seem possible to calculate both investment decisions and very realistic power flows in substantial detail. To the best knowledge of the author, no European long-term scenario published recently, applying a GEP model, features an explicit consideration of voltage levels. Consequently, for the research question under consideration a transshipment model seems be the best trade-off between accuracy and computational complexity. The implications for the modelling results will be discussed in chapter 4.

3.5.5. Time step

The temporal resolution is a central characteristic of electricity market models. While top-down approaches focus on yearly or monthly equilibria, bottom-up models typically have a much higher temporal resolution. The most common interval used for unit commitment is hours, for two reasons. Firstly, as most spot market contracts are traded in hourly products, the interval is very relevant for a realistic representation of real-world problems. The determination of power plant dispatch on an hourly basis follows the rules of the electricity market. Secondly, hourly data is readily available, being published by most exchange and grid operators.

Nevertheless, one could argue that a quarter-hourly resolution could increase realism. The need for quarter-hourly resolution increases with an increasing share of fluctuating RES-E and inflexibilities of the existing power plant park. In a system with low or no fluctuating generation, plant flexibility is decisive mostly to follow demand, in case of significant load forecasting errors or system faults. The gradient of the aggregated system load is usually not steep enough to cause significant problems for the power plant park. The ramp rate of the residual load to be supplied from dispatchable power plants becomes steeper through the diffusion of fluctuating renewable energies (Dallinger 2012). This creates a need for flexible power plants and might increase the necessity to model plant behaviour on a 15-minutes (or less) basis. Still, to the best knowledge of the author, no research has been published for theEuropean electricity market so far on a quarter-hourly basis. This is most likely because of the lack of publicly available data: for example, quarter-hourly wind power feed-in profiles are available only for a limited number of countries.Theoretically, some bottom-up modelling approaches, e.g. ACE, can calculate hourly market results and subsequently perform a re-dispatch of power plants. In this approach, the modelled generation companies could determine the optimal production schedule for their respective power plant fleet, taking into account hourly market results and quarter-hourly technical restrictions.

3.5.6. Time coverage

Another very central aspect concerns time coverage, i.e. the number of time steps covered in each model run. Here, a differentiation is necessary between UCM models and GEP models. UCM models often simulate each hour of the years in question. This means that the calculations are performed successively, simulating one day after another. This is not the case for GEP models, for which investment decisions have to be based on a large dataset consisting of knowledge or assumptions about future developments. Therefore, GEP models usually apply the *type day approach* and in rare cases *type weeks* (e.g. Schaber et al. (2012)). This means that for some exogenous data, e.g. for demand or meteorology, typical values are selected, and are intended to represent the remaining days of the year. Typically, demand is represented by one weekday, one Saturday and one Sunday/holiday. This data is provided for different seasons, which can either be winter and summer only, or can include all seasons. The model applied in Lienert and Lochner (2012), using a time step of two hours, observed 144 typical load levels, and is an example of a rather detailed time coverage. Type days are used in GEP models since basing investment decisions on all hours of a long time horizon exceeds the typically available computational resources. In some cases this issue is dealt with by calculating unit commitment for all hours of the year and deriving the investment decision through feedback-loops (Nicolosi, 2012).

Still, while the type day approach is well established for representing demand, it faces problems when being applied to multi-national power systems with differing holidays: it prohibits the integration of interactions between a country in which a certain day is a holiday with low electricity demand and its neighbours. The type day concept seems even less legitimate for reproducing the behaviour of fluctuating generation from RES. It becomes problematic in the following two cases:

1. *The model covers a large region:* For small, homogeneous regions it seems possible to define a set of profiles describing the characteristic behaviour of fluctuating generation from RES. For example, the generation from photovoltaics can be described exemplary for both a sunny and a cloudy day. However, problems arise, if correlations exist between different regions, which is for example the case for European wind regimes (Korpås et al., 2007, pp. 51-53). Distributing wind power plants over a large spatial area has a smoothing effect on the feed-in profile (IEA Wind Task 25, 2009; Giebel, 2005). Furthermore, existing seasonal correlations between wind and system load (Heide et al., 2010) can only partially be integrated. These effects cannot be included in

the type day approach, thus leading to false assessments regarding the necessity for transmission grids, storage facilities, a flexible generation park and its utilisation.

2. *The models covers multiple RES:* If the simulation deals with more than one fluctuating renewable energy source, correlations between these are likely. For example, negative correlations exist between wind and solar power in Northern Europe, as demonstrated by Widén (2011) for Sweden. Even if the type days are well selected, the correlations between between wind speeds and irradiation cannot be properly included.

The impact of the type days selected to represent the meteorology in the model grows with an increasing proportion of renewable energies. While for many countries the approach seems justifiable for today's power system with relatively low shares of fluctuating generation from RES, it becomes questionable for systems with a large share of renewable energy. Virtually all scenarios published in recent years dealing with an electricity sector complying with the 2°C target propose a RES-E share of 55 % or higher in 2050[12], demonstrating the high expectations regarding the growth of RES-E. For high shares of generation from fluctuating RES. The selection of the type days becomes crucial for the results where there is a high proportion of generation from fluctuating RES; a favourable profile leads to disregarding real-life issues. For example, in an optimisation model installing capacities in a certain site, region or country will seem overly beneficial if the chosen type days correlate well with the load profile or is negatively correlated to the profiles of the neighbouring regions. In turn, selecting unfavourable type days disregards the benefits that certain sites can bring in hours not represented in the type days. However, even GEP model used for scenarios with strong growth renewable energies, like the one applied in energynautics and EWI (2011) or Lienert and Lochner (2012) apply the type day approach. Research using hourly feed-in profiles for a whole year in GEP models have so far been restricted to one region or country (e.g. Nicolosi (2012)).

The type day approach assumes a typical, repetitive behaviour that de facto does not exist. Therefore, in scenarios with high proportions of RES-E, central generation technologies are depicted in a way that could be overly simplified. This is especially problematic as the fluctuating generation has a significant influence on other infrastructures, such as storages and electricity grids.In the current discussions, increasing the numbers of electricity storage facilities is often discussed as a central requirement for the integration of RE. The optimal level of storage facilities

---

[12]An overview and comparison of these studies is given in chapter 6.

and grids is strongly correlated with the utilisation of these infrastructures throughout their lifetime; costs and benefits cannot be assessed by sequencing the same, stylised situations. Consequentially, the investments concluded as being economic by the models (either for certain market participants or from a welfare point of view) could be significantly off the mark. Furthermore, applying type days limits the potential benefits of exchanging power between the nodes of the system, although this could be essential if the system is based on renewable energies.

Therefore, a central motivation behind this thesis i s to replace the type day concept by an approach to more accurately describe the characteristics of fluctuating RES-E. It is assumed that the accuracy increases with the time coverage. A substantial number of simulated weeks is assumed to be the minimum requirement for this work. At best, all hours of each considered year should be analysed based on realistic feed-in profiles for generation from fluctuating RES. The profiles could be taken either from published feed-in profiles or modelled from meteorological data.

3.5.7. Players' market behaviour

Electricity market modelling approaches can be differentiated between models with a single generator and models that have several market participants that compete against each other. Single generator models are typically optimisation models, in which the generator searches for the least-cost way of meeting a certain demand. This can also mean that all generators in the modelled region are aggregated in one generator. The optimisation can then be interpreted as the result of a market with perfect competition and does not represent a monopoly situation. [13] In turn, markets including two or more market actors with competitive strategies are analysed by equilibrium and simulation models. Of all characteristics presented in this section, the approaches used to represent the behaviour of market participants are most the heterogeneous in scientific literature. To present all concepts is beyond the scope of this work. Several publications either focus on classifying these strategies or include detailed summaries of existing approaches. For agent-based models, both Sensfuß (2007) and Weidlich and Veit (2008) present very detailed summary of existing approaches. In the case of equilibrium models, Ehrenmann and Smeers (2011) discuss a large variety of formal descriptions for strategic behaviour, focussing on investment decisions under uncertainty.

---

[13]Monopolies are hardly ever analysed for electricity markets due to the price in-elasticity of the consumers and the lack of realism.

For the research objective implementing competitive strategies is difficult for two reasons. First of all, in the large geographic area that has to be covered a very high number of firms are active in power generation. Platts World Electric Power Plant database lists 331 companies owning conventional power plants with generation capacities over 500 MW in Europe.[14] Although this number can be reduced by grouping affiliated companies, depicting the interactions between such a high number of players is very challenging for the computational complexity. Secondly, up until 2050, significant changes in the active market participants seem very likely. Processes of new entrants and the disappearing of incumbents are the subject of methodical modelling approaches, but is not often used in application-oriented models.

In consequence, the model to be applied in this work will be based on a perfect market and not consider explicit competitive strategies. Although this approach decreases the realism of the mid-term results, it increases transparency and minimises the need to speculate on strategies and interactions that cannot be based on fundamental data. Nevertheless, the implications of this decision for interpreting the results are discussed in chapter 4 .

3.5.8. Uncertainty modelling

The degree of uncertainty in the models varies substantially between the models. For approaches with no or very low uncertainty, one has to differentiate between approaches using perfect foresight versus step-wise modelling (Krey, 2006). In perfect-foresight modelling, the future developments of the variables is not only certain, but also known to the market actors or the model. In step-wise or myopic modelling, the future is modelled in steps, between which no endogenous linkages exits. For an optimisation model this means that although future developments are already certain, the optimisation has a short-term horizon. Myopic modelling is used either when there are computational limitations, or to "mimic" a limited foresight horizon of the market participants. Issues arise from the fact that in the case of myopic modelling, the players often have a perfect knowledge of the short-term future, but have no foresight of the mid- and long-term future. Therefore, the model is likely to base decisions on short-term events or artifacts.

The main reason for moving to stochastic modelling is that the derived strategies of market actor are more robust to changes or deviations from the assumed or most likely outcome (ibid.). Several exogenous and endogenous data can be the source of

---

[14]In total, the database contains 4,989 companies owning conventional power generation assets.

uncertainty. Variables that are typically modelled stochastically and examples for them include:

Fuel and carbon prices (Weber and Swider, 2004; Yang et al., 2008; Fuss and Szolgayová, 2010; Oda and Akimoto, 2011; Fuss, Szolgayová, et al., 2012): These parameters are often modelled stochastically, as their development is hard to forecast precisely and the market can be very sensitive to changes. An unforeseen change in fuel price can lead to investment decisions becoming unprofitable. The financial impact of false investment decisions is aggravated by the long life-time of electricity infrastructure. Therefore, especially single-firm GEP optimisation models often take uncertainties in this field into account. In recent research, uncertainties regarding the carbon price have become another focus of the modelling attention (Fuss, Szolgayová, et al., 2012). Although the effects and applied modelling approaches for carbon price uncertainties resemble the ones used for fuel prices, the possible price range for carbon allowances is often greater due to lower levels of empirical evidence and the influence of political decisions.

Demand (Kettunen et al., 2010; Kazempour and Conejo, 2012): Uncertain demand is again especially relevant for single-firm optimisation models. In this case, investment and bidding strategies have to be performed in order to maximise profit expectancy.

Competitors' strategies (Conejo et al., 2002; Plazas et al., 2005; Baíllo, 2002): In many models analysing strategies and market power the players do not know the strategies of their competitors. Therefore, they either have to maximise the expected profit of their strategy as a function of the other players actions, or apply learning algorithms to find the optimal strategy. [15]

Other variables are described stochastically in scientific publications. Messner et al. (1996) vary the specific investments of generation technologies. The approach used in Stein-Erik Fleten et al. (2002) models the inflows into the Nordic hydropower system as stochastic. The model developed in Sensfuß (2007) takes the availability of power plants into account stochastically. Furthermore, in models with exogenous power prices these are often modelled under uncertainty (Rajamaran et al., 2001; Stein-Erik Fleten et al., 2002). However, as these are only relevant for single-firm optimisation models, they are irrelevant to this thesis. The modelling approach

---

[15] As already explained above, a combination of the two, e.g. the ε-greedy-strategy, is also possible.

comparison performed in Ventosa et al. (2005) includes a description of how uncertainty is handled in the models presented. Similarly, a comparison of stochastic optimisation approaches is presented in Krey (2006).

Nevertheless it has to be mentioned that in order to model a spatially large electricity system over a long time horizon, stochastic modelling is rather uncommon. To the best knowledge of the author, none of the approaches used to analysed European long-term scenario studies applies uncertainty in the models. Presumably, the reason lies in the high computational requirements of the task itself, which multiplies if ranges have to be included for several parameters at the same time. Consequently, it is expected that uncertainty cannot be applied in the model used for this thesis. Nevertheless the benefits of stochastic modelling are recognised and strategies have to be defied for how uncertainty could be included in the model in the future. The model should be able to allow perfect foresight, as the changing market environment and necessary technological change is likely to render myopic investment decisions shortsighted.

3.6. Applicability of existing models

The previous section describes model capabilities and derives requirements for the model to be applied in this thesis. As already mentioned, a wide range of models has been developed in recent years with different foci. If a model existed that met the requirements, its application would be preferred in order to answer the research question. However, assuming that it is unlikely that any model will meet all requirements exactly, access to the source code of existing models, in order to perform the necessary modifications, is mandatory. This does not mean that the model has to be open source but also includes models for which the source code is commercially available.

Connolly et al. (2010) perform a extensive analysis of computer tools for analysing the integration of renewable energies. Not all of the 37 tools covered are electricity market models; of the electricity market models discussed, eleven are GEP models, which are presented in table 3.2. In the table, the *COMPOSE* model is included for the sake of completeness, but not taken into further consideration, as it deals only with single projects. The following section does not aim at a thorough description of the models but at screening the appropriateness of these models for the research question being considered under the conditions defined in section 3.5. The analysis relies mostly to on the information provided by Connolly et al. (2010), as their coverage often exceeds the publicly available material, as well as additional information by the models' operating companies or institutes.

Some of the available models are not applicable for the task at hand as they do not match the requirements defined in the previous section. These models will be presented first:

- *EnergyPLAN* is a deterministic Input-Output model developed by the Aalborg University, Denmark, covering the electricity market as well as the heat, transport and industry sector. Due to the lack of technological detail of top-down approaches explained in section 3.3, the model is not applicable for use in this research.

- *HOMER* model was developed by National Renewable Energy Laboratory in Golden, Colorado. The model can be used for simulating (and to a certain degree optimisation of) small energy systems. Although the model focuses on the effects of fluctuating generation, the model is designed only for small stand-alone electricity systems.

- *ProdRisk* is an optimisation model focussing on hydrothermal power generation, distributed by the Norwegian institute SINTEF. So far, the analyses carried out with the model cover electricity systems much smaller than the European system. So far, the analyses carried out with the model cover electricity systems much smaller than the European system. This and the fact that the model has so far focussed strongly on work on GEP advise against using this model.

- *ORCED* (Oak Ridge Competitive Electricity Dispatch) was developed by the Oak Ridge National Laboratory in the USA. The model uses an equilibrium-base approach. This model uses an equilibrium-based approach. The model's time horizon extending until 2030 and the temporal resolution, which is restricted to seasons, render an application to the task at hand impossible.

- The *PRIMES* energy system model was developed though several research programmes of the EU and maintained by the University of Athens. As the model is discussed in greater detail in chapter 6, it is not discussed here. The model is not available freely and is applied only by the University Athens for internal or contracted research.

Besides these models that can be disregarded, six models exist that match the requirement and are therefore taken into a more detailed consideration.

TRNSYS (TRaNsient SYstem Simulation) is a modular simulation model and is commercially available from a U.S., French and German consortium of institutes and companies. The model's source code is also available. TRNSYS has many features that could be beneficial for the research of this work, such as variable time steps and the option of integrating parts of the heat sector into the simulation. The model has been used to analyse systems with very high proportions of renewable energies. However, so far the model has been used to simulate rather small systems (Connolly et al., 2010). The model documentation[16] of the model shows that it is controlled mostly via a graphical user interface (GUI), which is unsuited for modelling a large system with several thousand power plants. Therefore, using TRNSYS is discarded as an option for carrying out the calculations for this thesis.

Mesap (Modular Energy-System Analysis and Planning Environment) *PlaNet* (Planning Network) is a modular model toolbox. The framework can be adjusted to different model outlays and the spatial and temporal resolution match the requirements. Furthermore, the model has already been applied to similar research, most prominently the modelling for Greenpeace's *energy [r]evolution* scenarios (Teske, A. Zervos, et al., 2007). The model is commercially available, and the fee is in an acceptable range. However, after a discussion with the scientific group maintaining and distributing the model, its application for the work of this thesis was discarded: although the model environment generally seems very suitable, it could not be determined whether the internal handling of the model data would have been fast enough for the massive flows of data needed for the given task. Despite the possibility that it would have been fast enough, it would only have become clear once the model was (nearly) complete. A potential failure at such a late point is regarded as too risky.

EMCAS (Electricity Market Complex Adaptive Systems) was developed in 2002 by Argonne National Laboratory, who maintain and distribute the model. Although the model was initially a UCM model, it was expanded in 2007 to include GEP. The model uses an agent-based approach. The user is given the option of defining the agents' objectives in high detail and include endogenous learning. The electricity grid is modelled applying a DC flow approach. The main obstacle for applying EMCAS for the task at hand is calculation time. The available description (EMCAS, 2008) states that the time required for a simulation with 10 nodes and 70 power plants is one hour on a 2.0 GHz processor. Applying the model to a scenario with several thousand power plants is likely to result in very long simulation runs.

---

[16]A short description of the model and its application is available at the models website (TRNSYS, 2011).

Whether an agent-based GEP market model with detailed modelling of the grid can be handled for a large region over a long time horizon is doubtful. To the knowledge of the author, such a modelling exercise has yet to be executed successfully. Furthermore, as the source code of the model is not available, it is uncertain if the agent-based decision processes can be successfully adopted to systems with with very high proportions of renewable energy.

BALMOREL is an optimisation model covering the electricity and district heating system. The model is developed and maintained as open source written in the GAMS programming language. The defined optimisation problem can be solved with different solver programs, which can be open source, such as GLPK, or commercial programmes, such as CPLEX.[17]. Data management is usually carried out by Microsoft Office products such as Microsoft Excel and Microsoft Access. As the model is open-source and can be modified, several versions of it exist and have been developed in parallel. The time horizon, temporal resolution and the region covered can be freely defined. Grid aspects are handled through a transshipment model. In summary, the model seems suited for carrying out an analysis such as that envisaged in this work. However, the experience of the model with scenarios incorporating high proportions of RES-E and with the optimisation of large regions is rather small. Also, according to Connolly et al. (2010), when all hours of the year are calculated, the time horizon is usually limited to one year. Furthermore, in the experience of the author, Microsoft Access has certain weaknesses when handling large amounts of data. Other database management software, such as SQL (Structured Query Language), is often superior in terms of speed and stability.

TIMES (The Integrated MARKAL-EFOM System) is an optimisation model based on the MARKAL model, the two forming a model family often referred to as MARKAL/ TIMES. The model was developed and is maintained by the International Energy Agency (IEA). The model is available as open-source, with many institutes and companies worldwide applying the model and adjusting it to specific research questions. The model is similar to BALMOREL as it is also implemented in GAMS with different solvers to which it can be linked. TIMES covers various energy forms and sectors, but can also be applied solely to the electricity sector. Energy demand can be modelled price elastically. Furthermore, some versions of the model can be linked to macroeconomic or environmental modules. The model can be applied to a large range of regions, from community to world level. As in most optimisation models, demand is exogenous, for which the model determines

---

[17]The solver is discussed in chapter 4.

the least-cost supply. The temporal resolution can be adjusted, but is typically below 100 time steps representing a year. The fact that no experience is documented for the model's run-time in a very high temporal resolution is an obstacle for its application in this work.

PERSEUS (Programme-package for Emission Reduction Strategies in Energy Use and Supply-Certificate Trading) was developed by the Institute for Industrial Production at the University of Karlsruhe and the Brandenburg University of Technology Cottbus, who also maintain the model. The model is mainly sold to large utilities or used for research within the universities. The model is based on an optimisation approach with a strong emphasis on stock flows (described in Fichtner (1999)) and emission trading (described in Enzensberger (2003)). The model has been used to analyse the effects of the diffusion of renewable energies in the EU-15 (Rosen, 2008). The optimisation uses a type day approach, which is restricted to 26 to 72 time steps per year. The model's source code is not available.

It has to be noted that numerous other models exist that are proprietary, i.e. owned and run only by a certain company or institute. As these models are not freely available, they are not discussed here. Features and methods of these models that have been made public through scientific publications or model documents and are relevant for the questions being considered are presented in the respective sections of 4.

3.7. Conclusion and critical reflection

This chapter introduced the main concepts and approaches used in electricity market and system modelling. The comparison shows that no concept is dominant, but all have their particular strengths and weaknesses. Electricity market models focus on the effects of imperfect markets as well as players' strategies and interdependencies. While the models deliver realistic results, especially in short- to mid-term horizons, they come with certain limitations in terms of size of analysed region and time horizon. Electricity system models in turn feature a less detailed view on realistic market processes, but allow for the analysis of larger systems and longer time horizons.

The capabilities of the model that will be used for this thesis were defined, and existing models assessed on this basis. The main conclusion is that in the face of the growing proportion of fluctuating generation, a higher temporal resolution seems necessary, especially in terms of time steps covered by the models. The assessment revealed that no existing model is designed to model a long-term scenario of the

European electricity market with a very high time coverage. Where an existing model is theoretically capable of hourly resolution throughout the entire period, the consequences in terms of the model's run-time are uncertain. The fact that, to the knowledge of the author, none of the models has been applied to such a task seems to indicate that the run-times of the models increase to levels that are not manageable for scientific research. Even if a model is able to produce a result within a couple of weeks, the long run-time would significantly limit the options for necessary sensitivity analyses.

For this thesis it means that two options exist for creating a suitable model. The first approach is to create a completely new model from scratch. While this approach has the advantage that the model can be exactly customised, it is challenging as all components have to be developed. This includes the tasks of data management and evaluation. The second option is to alter an existing model to tailor it to the requirements. The disadvantage of such an approach is that the suitability of a particular model cannot be ascertained in advance. If during the development process it transpires that a certain requirement cannot be met, the work up to this point is in vain. The main obstacles in this context are again run-time and data management. Another issue for adapting an existing model is the level of technological detail. Although all of the models presented above represent conventional power plants with a high level of detail, the concepts for modelling renewable energy technologies are not always transparently documented. This can become problematic for technologies such as reservoir hydropower plants or concentrated solar power (CSP), for which the generation behaviour is not entirely dominated by the supply of the respective renewable energy. It is difficult to assess whether or not the specifications of these technologies can be integrated into existing models before starting development.

To avoid the risk of a failure to adopt a model in some stage of the development process, a mixture of both options described above is chosen. The modelling in this work builds on the existing electricity market model PowerACE, one version of which is maintained by Fraunhofer Institute for Systems and Innovation Research ISI, the institute employing the author during the research of this thesis. The model in its current version is a UCM model focussing on the German power sector, applying an agent-based approach. As noted earlier, all models used for long-term scenarios of the European electricity sector with a high level of technical detail are optimisation models. With currently available computers, only the simplified market representation of optimisation models allows the calculation of GEP covering large regions with high temporal resolution by using well-established solver programmes. Consequently, for creating a model for the task in this thesis, Power-ACE is stripped bare of its core, agent-based modelling concepts. Only the data management, which is well-established, adaptable and fast is kept. The main parts of the model, i.e. the GEP and UCM components, are re-developed from scratch as an optimisation model.

This approach has several benefits. First of all, it provides for exact tailoring of the model to the envisaged task. The model can be designed explicitly to allow an hourly temporal resolution throughout the year. In the programming process, computing speed and run-time can be made priority objectives. Secondly, the approach relies on an existing, transparent and sufficiently tested model for data handling. The PowerACE model possesses databases and efficient loading structures for most data necessary for the required task and additional databases can be integrated quickly. Thirdly, this approach has the advantage that it provides for a future multi-approach platform. Although this is beyond the scope of this work, the results of the model to be developed, for example in terms of future power plant and transmission capacities, could be fed into the PowerACE UCM market model to evaluate and compare the results. This would, to some degree, combine the strengths of market and system models.

The decision to apply an optimisation has certain disadvantages however, the most significant of which being a relatively weak representation of real-world market processes. The strategies of market participants as well as pricing mechanisms can only be incorporated in a very simplified manner. Nevertheless, this weakness is less grave in long-term scenarios, in which the rules of markets and the strategies applied are deemed to change significantly. An optimisation approach prevents the need to speculate on these rules and strategies and delivers transparent and comprehensible results. Comparing or supplementing results, especially for the early years for which the market rules are known, with the results of models with a more realistic market representation is regarded as a valuable point for future research.

To the best knowledge of the author, a long-term modelling of the whole European power sector with a high temporal resolution and coverage is a novelty. Integrating these aspects accounts for the central role that fluctuating generation will presumably play in the electricity system and the challenges arising from this development.

Table 3.2.: Existing GEP electricity market models taken into consideration. Source: Connolly et al. (2010)and own research.

Model	Type	Application	Geographic area	Time step	Horizon	Availability
COMPOSE	optimisation	Project assessment	Single project	No limit	Hourly	Free Download
EnergyPLAN	Input-Output	National Energy System	National/regional	Year	One year	Free Download
HOMER	optimisation	Stand-alone system	Local/community	Minutes	One year	Free Download
ProdRisk	Optimisation	Hydropower	Local/regional	Hours	Years	Commercial
ORCED	Equilibrium		National/regional	Hours	One year	Free Download
PRIMES	Equilibrium	Intersectoral equilibriums	National/regional	TDH^a	50 years	Commercial
TRNSYS	Simulation	Community systems	Local/community	Seconds	Years	Commercial
Mesap Planet	Modular	National energy systems	National/regional	Any	No limit	Commercial
EMCAS	ACE		National/regional	Hourly	Any	Commercial
BALMOREL	Optimisation		International	TDH	50 years	Open source
TIMES	Optimisation		International	TDH	Unlimited	Open source
PERSEUS	Optimisation	Several applications	International	TDH	50 years	Commercial

[a] TDH: Type days are modelled in hourly resolution.

4. Development of the model PowerACE-Europe

The previous chapter defined the necessary capabilities for the model to be developed and applied in this thesis. The main objective is to create a model that calculates both capacity expansion and unit dispatch for Europe with high temporal resolution and coverage. A strong emphasis has to be placed on the realistic representation of the characteristics of electricity generation from RES.

This chapter describes the main components of the resulting model *PowerACE-Europe*, the model developed for calculating the scenarios analysed in this work. The description starts in section 4.1 with a brief introduction to the relevant existing parts of the PowerACE model cluster which are used and in some cases adapted. The biggest alterations to existing components are summarised in section 4.2. The new components of the model and the functioning of the utilised solver software are discussed in sections 4.3 and 4.4, respectively. The chapter concludes in section 4.5 with a critical reflection on the functioning of the model and its capabilities as well as options for further developments.

As a convention, the new model developed and described in this work will henceforth be called PowerACE-Europe. The name is slightly misleading: unlike the existing PowerACE model, which is agent-based, the new version of the model does not contain agent-based components. The name is chosen purely because the model is part of the *PowerACE model cluster* and should be seen in the context of the model family within which it is embedded.

4.1. The PowerACE model cluster

PowerACE-Europe is embedded in the existing PowerACE cluster system. The version on which it is based focuses on an agent-based power market model of the German power markets. A detailed description off the model is given by Sensfuß (2007) PowerACE seeks to simulate the markets in several modules, which are implemented in the programming language *JAVA*. The central model of the cluster, the *PowerACE simulation model*, focuses on the matching of demand and supply on both spot and reserve markets. The first version of the model was originally developed in cooperation between Fraunhofer Institute for Systems and Innovation

Research ISI, the University of Karlsruhe and the University of Mannheim and was sponsored by the "Volkswagen Stiftung". In its current version, it also deals with other aspects of the power market, such as electric mobility and the retail market. A simplified visualisation of the structure of the central part of the simulation model is depicted in figure 4.1. The model has a high level of technical detail and market rules. The agents have a restricted access to information on the market, which they use to act within their means. The simulation follows the rules of the central German power markets, i.e. the spot markets as well as the primary and secondary reserve market. The generation side of the model includes generation from conventional power plants and RES, as well as production from pumped storage hydro power plants.

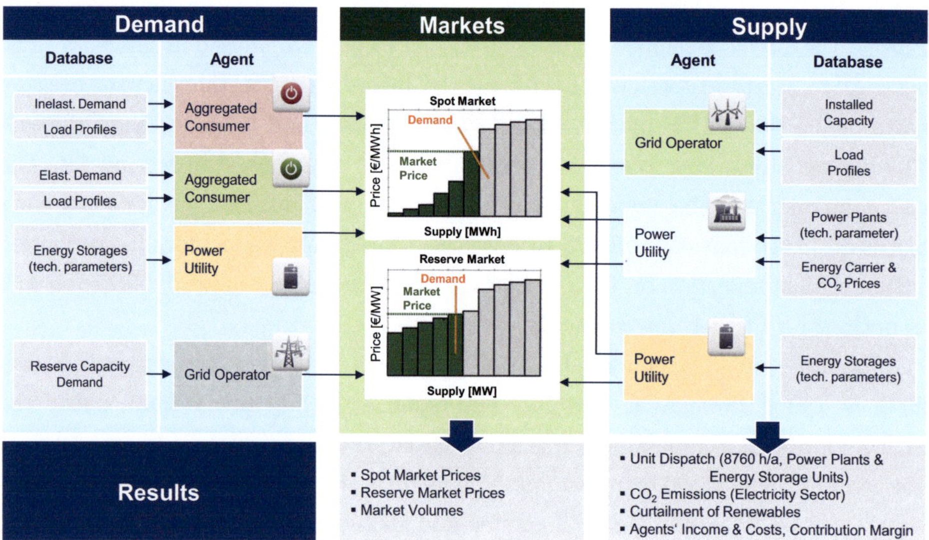

Figure 4.1.: Simplified structure of the PowerACE simulation model, based on F. Genoese, M. Genoese, et al. (2012).

4.2. Changes in the common modules of the PowerACE model

Since only a relatively small proportion of the modules of PowerACE is actually used in the development of PowerACE-Europe, a detailed description of the model is not provided here.[1] In the following, only the components which have been significantly altered by the author of this thesis are explained.

---

[1] A detailed in-depth description of the model and its functionalities is given by Sensfuß (2007).

The source code of the whole PowerACE model cluster is stored in a subversion system, which automatically tracks all changes to it. This means that the code of the model contains the source files for both the agent-based simulation model and the newly developed optimisation model. Which of the two approaches the user chooses to apply is controlled via a single switch variable. During the execution of the programmes, the initial steps in the model, most of which concern data management procedures, are identical for both variants. At a certain point in the model execution, the two approaches follow different paths.

All data used by the model is stored in $MySQL^2$ databases located on a server at the Fraunhofer ISI. The databases store both input data, e.g. scenario specific data on demand and power plants, as well as general model control parameters. The latter ones can be set manually via a front-end or by the PowerACE component *Scenario Creator*, which provides a structured GUI for defining and managing scenario parameters.

The original PowerACE model only deals with the German electricity sector. Electricity trade between Germany and the neighbouring countries is considered to be exogenous. The hourly import and export of electricity is stored in a database, containing data on commercial flows of electricity, published by grid operators (see: ENTSO-E (2012a)). In the model, the international trade is "forced" into the market: export is represented as demand bids at the maximum bid price, whereas imports are represented as supply bids at the minimum price. The electricity grid and related restrictions within Germany are assumed to be irrelevant for the spot market, meaning that the model applies a single-node grid model. Therefore, for the model to be able to cover Europe it is necessary to convert its architecture and data management into a multi-regional structure.

The regional set-up is controlled through several variables and a scenario database allowing the user to define the regions to be taken into consideration in the model runs. These regions do not necessarily have to be countries, as a higher or lower resolution can be applied. In the work at hand, only the EU-27, Norway and Switzerland are considered, each country represented as an individual region. Due to the new spatial and temporal differentiation of the model, most databases of the model had to be restructured in order to assign the data to the respective regions. In all databases linked to the multi-regional components, the new spatial dimension is included as a column, as exemplified in table 4.1. Although the details are of low scientific interest, it should be acknowledged that these changes required

---

[2]MySQL is a relational database management system. The programme is open source and available at www.mysql.com.

a substantial alteration of the data management structures in both the databases and the model itself. The addition of a spatial variable requires values that had previously been stored in a table-like format to be restructured into other, more database-conform multi-identifier formats. As these changes affect virtually all data stored and used in the model, the task required the adaptation of several thousand lines of code.

Table 4.1.: Exemplary changes in the databases through the introduction of a spatial dimension *region*.

Scenario	Region	Year	Hour	Electricity demand
Baseline	Germany	2020	0	65.000 GW
Baseline	Germany		1	66.500 GW
⋮	⋮	⋮	⋮	⋮
Baseline	Germany	2020	8759	68.200 GW
Baseline	France	2020	0	51.900 GW
Baseline	France	2020	1	54.200 GW
⋮	⋮	⋮	⋮	⋮

The new spatial differentiation developed for the optimisation model can also be applied to the existing agent-based version of PowerACE. This required changes in the way the spot and reserves markets are treated in the model as well as the way the agents submit their bids to these markets. As the scenarios calculated for the subsequent work only apply the optimisation model, these changes are not discussed here. However, this is the first step and prerequisite for a future hybrid version of the model.

The original model calculates the results myopically for one year. For PowerACE-Europe, two options are possible. The model can be either be set to myopic mode, in which case only one year is calculated. Alternatively, the model can be run in perfect foresight mode, in which the linear problem is defined for several *scenario years*, which are connected through inter-annual linkages.[3] In this case, a number of years can be defined to be calculated explicitly, and these do not necessarily have to be consecutive. The number of years is limited only by the available memory of the computer. In all scenarios presented in this work, the years 2020, 2030, 2040 and 2050 are calculated in perfect foresight. The ability to calculate several years

---

[3]The nature of the inter-temporal connections between the years is explained in the respective sections, as it varies for the different system components.

in an integrated manner is currently not applicable to the agent-based version, as it lacks procedures for simulating capacity expansion.

4.3. Components of the newly developed optimisation model PowerACE-Europe

The functioning of the model PowerACE-Europe can be summarised as follows. The model starts by loading all relevant data, mostly using the same methods as other PowerACE components. Then a formal description of the optimisation problem is generated, passing the resulting linear problem to the solver software. The solver calculates the optimal solution, which is then processed by the model and logged to files or databases. Since the optimisation problem is rather large and several modules of the model participate in its formal definition, the description in this thesis is structured in a similar way, explaining components of the model and their contribution to the linear problem.

4.3.1. The Linear Problem Manager

The Linear problem manager (LPM) is the central component of the optimisation model. All input data and parameters used in the model enter the LPM directly or in a processed form. The task of the LPM is to handle the linear programme (LP) describing the scenario and the communication with the solver software. For the work at hand, the solver is the *IBM ILOG CPLEX Optimizer*, referred to in the following as CPLEX. A short description of the software and the algorithms used is presented in section 4.4. The LPM is also used to control the parameters of the solver software, such as the applied solver algorithm, required solution accuracy, maximum calculation time and the number of processors or threads[4] available for the solver.

Each linear problem consists of an objective function, and restrictions, which are referred to as bounds. In standard form, an LP can be formulated as:

$$
\begin{aligned}
\underset{x}{\text{minimise}} \quad & c^T x \\
\text{subject to} \quad & Ax \leq b \\
\text{and} \quad & x \geq 0
\end{aligned}
$$

---

[4]Some solver algorithms apply *multithreading*, allowing the computer's Central Processing Units (CPU) to work in parallel on the LP.

The objective function of the LP $c^T x$ set up by PowerACE-Europe contains all monetary costs c linked to the the decision variable vector x attributed to the time covered by the model. This means, that the objective function contains:

- capital costs,

- fixed costs for operation and maintenance (O&M),

- variable O&M costs,

- fuel expenditures and

- costs of acquiring ETS emissions allowances, i.e. "CO_2-costs"

The capital costs are annualised and taken into account only for the scenario years. Depending on the applied discounting procedure, all costs have to be discounted to one year, which changes the weights between the years.[5] Furthermore, in some cases *virtual costs* are added to prevent the model from performing unwanted actions in otherwise ambiguous situations: The unwanted actions are penalised by attributing (very low) costs to it, which are not evaluated in the solution.

The bounds in the LP describe the setup of the scenario, with each forming one line. Most bounds are defined by one module alone, with the exception of the demand inequation, to which almost all modules contribute.

4.3.2. Conventional power plants

The Conventional Power Plant Manager (CPPM) module handles all variables related to power generation from conventional fuels. This applies to plants to be built in the future scenario years as well as existing power plants. In the case of the former, capacity expansions of a technology are associated with costs in the objective function. For existing power plants, the linear problem consists only of unit commitment aspects, i.e. finding the cost-efficient dispatch, because their fixed costs are considered sunk costs and therefore irrelevant for decision-making.

The existing power plants are stored in a database, the input data of which is largely based on Platts' World Electric Power Plant (WEPP) database (Platts, 2010). This database is commercially available and contains detailed data on electricity generating units worldwide. The following data is loaded by the model PowerACE:

---

[5] As this approach has significant implications, the issue is discussed in section 4.5.3.

- Name of the plant,

- location (i.e. country and in some cases also region and city),

- primary fuel,

- initial year of operation and

- gross rated power.

For the applicability in the model, this data has to be supplemented by data on net rated power, expected year of plant retirement and electric efficiency. To calculate net rated power CAP_p^{net} of a power plant p, the plants self-consumption rate $\varsigma(f_p)$ is subtracted from the gross rated power CAP_p^{gross}. The self-consumption rate is simplified to be a function of the respective fuel f_p of the power plant. It is set to 6 % for coal power plants, 7 % for lignite power plants and 5 % for all power plants utilizing other fuels, based on rounded averages of data published by various power plant operators.

$$\mathrm{CAP}_p^{net} = \big(1 - \varsigma(f_p)\big)\mathrm{CAP}_p^{gross} \tag{4.1}$$

with:

p	power plant index
f_p	fuel index
$\varsigma(f_p)$	plants self-consumption rate [1]
CAP_p^{net}	net rated power of power plant p [MW]
CAP_p^{gross}	gross rated power of power plant p [MW]

The expected lifetime of a power plant is assumed to be a function of the turbine type, which is available from the WEPP database. The assumed lifetimes are shown in table 4.2 and are based on data published by power plant operators as well the WEPP database, which includes data on retired plants. In the literature, a large range of assumptions on plant lifetimes can be found, often deviating from the ones applied here. For example, Ludig et al. (2011) assume a longer lifetime of 50 years for coal and lignite power plants and shorter lifetime of 30 years for gas. The possible influence of other settings for these parameters is discussed in section 6.3.

Lifetimes of individual plants are adjusted in the database if new information becomes available. This is for example the case for the nuclear power plants in

Table 4.2.: Expected lifetime of power plants depending on the turbine type.

Turbine type	Expected lifetime
Open-cycle gas/combustion turbine	40
Steam turbine	40
Combined-cycle gas turbine	40
Internal combustion engine	25
Fuel cell	20
Turboexpander or gas expander	20

Germany, for which the retirement years are set by individual legislation based on the nuclear phase out decided in "Bundestags-Drucksache 17/6246".

The electric efficiency of a power plant μ_p is assumed to be a function of its gross capacity, first year of operation INIT_p, fuel f_p and turbine type TURBINE_p.

$$\mu_p = f(\text{CAP}_p^{gross}, \text{INIT}_p, f_p, \text{TURBINE}_p) \tag{4.2}$$

with:

INIT_p Initial year of operation of power plant p

TURBINE_p turbine type of power plant p

For the first two parameters a weak positive correlation is assumed, i.e. the efficiency tends to increase the larger and newer a plant is. The initial year of operation serves as an input parameter as the efficiency usually increases (at least up to a certain point) with higher steam temperatures and pressures. Over time, the achievable steam parameters increase through new technical developments, e.g. cost-efficient materials able to cope with higher temperatures. The function is derived from a database in which known parameters are stored for the different fuels and turbine types. Between the data points the values are interpolated.[6]

The previous parts describe how data concerning power plants is handled in PowerACE in general. To apply the data in PowerACE-Europe, the power plants have to be aggregated. In the model, all power plants of a country c using the same fuel are summarised by their efficiency into a fixed number of aggregated plants $\hat{p}$. The reason for this lies in the computational resources; to include individual

---

[6]However, the weak positive correlation between size and initial year of operation is a simplification, since in the turbine market, like in most markets, a "budget segment" exists. A turbine installed in a certain year does not necessarily have to be of the best available technology but can also be a cheaper model with lower efficiency.

power plants would only bring substantial benefits to the model if the optimisation problem was changed to a mixed-integer formulation. This would allow, for example, additional model restrictions to increase power plant capacities discretely in typical plants sizes. Although MIP is theoretically possible for both PowerACE-Europe and CPLEX, it substantially increases the model's run-time, as tests with similar but smaller MIP models have shown. Consequentially, applying MIP is discarded, but might be taken up again once faster algorithms or computers are available or if the problem at hand is smaller, as may be the case , for example, if only a small number of countries is considered.[7]

The number of aggregated plants per country depends on the fuel type. Power plant types with a rather small range of typical efficiencies, such as nuclear power plants, are aggregated into a smaller number than power plants with a more heterogeneous efficiency, for example gas power plants. All power plants using the same fuel are sorted by their efficiency and divided into groups that are of a similar size in terms of cumulated net rated power. The aggregated power plants, indexed $\hat{p}$ use the same fuel as the original plants, their installed capacity equals the sum of the original plants and their efficiency is derived as an weighted average. The procedure is repeated for each simulation year.

$$\text{CAP}_{\hat{p}}^{gross} = \sum_{\hat{p}\epsilon\mathcal{P}} CAP_{\hat{p}}^{gross} \tag{4.3}$$

$$\mu_{\hat{p}} = \frac{\sum_{\hat{p}\epsilon\mathcal{P}}(CAP_{\hat{p}}^{gross}\mu_{\hat{p}})}{\sum_{\hat{p}\epsilon\mathcal{P}} CAP_{\hat{p}}^{gross}} \tag{4.4}$$

with:

$\hat{p}$ index of the aggregated power plant

$\mathcal{P}$ set of power plants belonging to aggregated power plant plant $\hat{p}$

The fact that power plants are considered in the model in an aggregated form has implications for the precision of the model, for example on its ability to describe partial load behaviour. Theoretically, partial load efficiency can be approximated by a linear model. Figure 4.2 exemplifies how a fictitious 100 MW power plant

---

[7]However, as the currently available resources are already very powerful, it cannot be expected that solving a problem similar to the one at hand with a single computer is possible in the near future. It might be worthwhile to investigate whether the use of cloud-computing can help tackling this restriction.

can be represented by three blocks. "Virtual power plant 1" is modelled with the efficiency assumed for the minimal load of 60 MW.[8] The other two virtual power plant blocks of 20 MW each have a higher efficiency used to generate the desired partial load behaviour depicted on the right side of figure 4.2. The utilisation of the virtual blocks is restricted in a way that they can only operate when the block below is already running at full capacity.

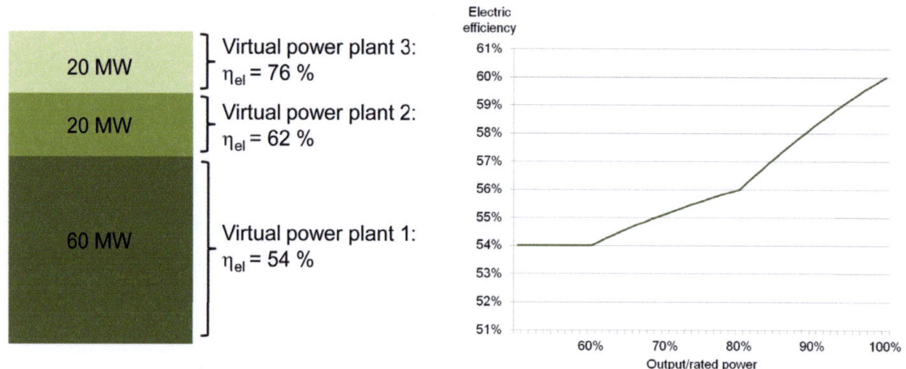

Figure 4.2.: Set-up of a fictitious 100 MW power plant (left side) and resulting partial load efficiency (right sight).

However, an aggregate of power plants, even if homogeneous in their technical parameters, does not necessarily behave like a larger plant of the same type. As an example, imagine five power plants of 100 MW net generation capacity each having to supply 400 MW of power. This could be achieved by each plant supplying 80 MW, consequently running in partial load below their rated efficiency. However, the load might also be supplied by shutting one plant down completely while the others operate at full power and efficiency. The actually realised dispatch depends on technical factors, for example, transmission constraints, and can also be influenced by the results of reserve markets, which may partially bind some power plants. However, if such restrictions do not occur, all power plants, besides the last one needed to meet demand, can be assumed to operate at rated power and efficiency (or sufficiently close to it). Consequently, the approximation procedure described above is not applied by PowerACE-Europe due to the current lack of robust data for parametrisation..

---

[8]In an MIP formulation, the utilisation of this block could be of a "boolean" type, meaning that the power plant could either be shut down or operate with at least 60 MW.

The aggregated power plants representing the historic power plant park are described formally by the CPPM, which hands the linear expressions to the LPM. As the power plants are already build, the model needs to calculate only the optimal utilisation of the plants. Therefore, the formal representation in the objective function, given by eqn. (4.5), consists only of the variable costs associated with the production of one MWh of electricity from the respective aggregate of power plants. Fixed costs are considered to be sunk costs and are not part of the optimisation model.[9] Eqn. (4.6) represents the technical limitation, whereas eqn. (4.7) implements the contribution of the plants' generation for meeting demand.

$$\text{minimise:}_{\vec{x}} \ \ldots + \sum_{a \in \mathcal{A}} \sum_{\forall h} \sum_{\forall c \in \mathcal{C}} \sum_{\forall \hat{p} \in \mathcal{P}_c} x_{ahc\hat{p}} \left(\underbrace{\frac{p_a^{f_{\hat{p}}} + p_a^{\text{CO}_2} \cdot \kappa_{f_{\hat{p}}}}{\mu_{\hat{p}}}}_{\text{fuel and emission costs}} + \text{O\&M}_{\hat{p}}^{var} \right) + \ldots \ (4.5)$$

subject to:

$$0 \leq \qquad x_{ahc\hat{p}} \qquad \leq \gamma_{\hat{p}} \cdot \text{CAP}_{\hat{p}}^{gross} \quad \forall a, h, c, \hat{p} \qquad\qquad (4.6)$$

$$[DS]:^* \ \ldots + x_{ahc\hat{p}} + \ldots \geq D_{ahc}^{res} \qquad\qquad \forall a, h, c, \hat{p} \qquad\qquad (4.7)$$

with:

a	year index
h	hour of the year index
c	country index
$f_{\hat{p}}$	fuel index of power plant $\hat{p}$
$\mathcal{A}$	set of scenario years, e.g. {2020, 2030, 2040, 2050}
$\mathcal{C}$	set of scenario countries
$\mathcal{P}_c$	set of all (aggregated) historic power plants located in country c
$x_{ahc\hat{p}}$	net electricity generated in year a in hour h in country c by power plant $\hat{p}$ [MWh$_{el}$]
$p_a^{f_{\hat{p}}}$	price of fuel used in power plant technology $\hat{p}$ [EUR]

---

[9]Naturally, this is a simplification: Fixed costs usually also include insurance costs and, at least partially, labour costs and certain taxes. These costs could be saved through closing down the power plant.

$p_a^{\mathrm{CO_2}}$ price of carbon allowances in year a [EUR/t]

κ_f CO_2 content of fuel f [$t_{\mathrm{CO_2}}/\mathrm{MWh}_{th}$]

$\mathrm{O\&M}_{\hat{p}}^{var}$ O&M costs in plant $\hat{p}$ for the generation of 1 MWh [$\mathrm{EUR/MWh}_{th}$]

$\gamma_{\hat{p}}$ statistic availability of plant $\hat{p}$ [1]

D_{ahc}^{res} residual electricity demand [10] of in year a in hour h in country [MWh]

* This inequation reflects the demand and supply of each country. Only the contribution from conventional, existing power plants is shown here. Throughout this chapter, the inequation is marked with *[D-S]*.

Capacity expansions of conventional power pants are handled in a similar way, but are subject to the implicit investment planing process. This means that the problem is formulated with additional variables representing the generation capacities to be installed in future years. In the following, by convention, all investment decision variables are indicated by a capital X, whereas all variables dealing with the (hourly) utilisation of infrastructures are indicated by x.

$$\text{minimise:} \quad \ldots + \underbrace{\sum_{a \in \mathcal{A}} \sum_{\forall \alpha \in \mathcal{A}_a} \sum_{\forall c \in \mathcal{C}} \sum_{\forall p \in \mathcal{P}} \left[\tau_{ap\alpha} \delta_a X_{cp\alpha} \mathrm{COST}_{p\alpha}^{fix} \right.}_{\text{discounted annualised fixed costs for investments}} \tag{4.8}$$

$$\underbrace{+ \sum_{\forall h} \delta_{ap\alpha} x_{ahcp\alpha} \left(\frac{p_{f_p a} + \kappa_{f_p} \left(p_a^{\mathrm{CO_2}}(1 - \mathrm{CR}_{\alpha p}) + p_a^{\mathrm{CCS}} \mathrm{CR}_{\alpha p} \right)}{\mu_p} + \mathrm{O\&M}_{p\alpha}^{var} \right) \right]}_{\text{variable costs}} + \ldots$$

subject to:

$$0 \leq \quad x_{ahcp\alpha} \quad \leq \gamma_p \cdot X_{cp\alpha} \quad \forall a, h, c, p \tag{4.9}$$

$$[DS] \ldots + x_{ahcp\alpha} + \ldots \geq D_{ahc}^{\mathrm{res}} \qquad \forall a, h, c, p \tag{4.10}$$

---

[10]Residual demand in this context refers to the net electricity demand including losses of the internal grid less exogenous RES-E (see section 4.3.6).

with:

p	power plant technology index
f_p	fuel index of power plant technology p
$\mathcal{P}$	set of available power plant technologies p
$\mathcal{A}_a$	sub-set of $\mathcal{A}$, containing all scenario year equal to or less than a
$\tau_{ap\alpha}$	binary variable: "1" if $a - \alpha <$ the plants lifetime, "0" otherwise.
δ_a	discounting factor for costs incurring in a
$X_{cp\alpha}$	investment decision variable, i.e. capacity increase in power plant technology p in year α in country c [MW]
$x_{ahcp\alpha}$	net electricity generated in year a in hour h in country c by power plant technology p built in year α [MWh$_{el}$]
$\text{COST}_{p\alpha}^{fix}$	annualised specific fixed cost of power plant technology p built in α [EUR/MW]
p_{f_p}	price of fuel used in power plant p [EUR/MWh$_{th}$]
p_a^{CCS}	carbon price in year a [EUR/t]
$\text{CR}_{\alpha p}$	Capture rate, i.e. share of emissions captured by technology p built in year α. Larger than 0 only for CCS technologies [1]

Through eqn. (4.9) the hourly power generation of each technology is restricted to be lower than the installed capacity, taking into account the statistic availability. Investments into power plants, described through $X_{cp\alpha}$, are linked with fixed costs, occurring if the models "decides" to increase the capacity of the respective technology. These costs consist of the annualised specific investments as well as the fixed O&M cost (see eqn. (4.11)). The annuity factor is calculated on the basis of the assumed interest rate, i.e. the weighted average cost of capital (WACC), and the lifetime of the power plant. Capital costs are distributed evenly over the plant's lifetime.

When defining the techno-economic parameters, the continuous chronology of the investment decisions has to be considered. These are simplified in the model to take place in discrete steps; the investment process described by the variable $X_{cp\alpha}$ takes place before the year α. The power plant capacity expansion calculated to be optimal for example for the year 2030 would have to be built between 2020 and 2029, to be available in the year 2030. The power plants' parameters have to be selected accordingly to represent an adequate average of the properties for the 2020 and 2029.

$$\text{COST}^{fix}_{p\alpha} = \underbrace{\frac{(1+i_p)^{n_p} \cdot i_p}{(1+i_p)^{n_p} - 1} I_{p\alpha}}_{\text{annualised investments}} + \text{O\&M}^{fix}_{p\alpha} \tag{4.11}$$

with:

i_p interest rate for technology p [1]

n_p lifetime of technology p [years]

$I_{p\alpha}$ specific investment of the power plant technology p built in year α [EUR/MW$_{el}$]

$\text{O\&M}^{fix}_{p\alpha}$ fixed O&M cost per year for power plant technology p built in year α [EUR/MW$_{el}$]

Although the linear equations describing the investment process are similar to those describing the utilisation of historic power plants, the effects in the model are very different. Investment decision variables are highly inter-temporal: a variable X_{cp2020}, representing a decision to increase the capacity installed in a certain power generation technology in 2020 in country c is linked to 70,080 equations or inequations directly and through these equations to virtually all decision variables in the model.

The utilisation of the future power plants also creates costs in the model's objective function, the description of which is very similar to those of existing power plants. However, depending on the scenario, power generation technologies can be equipped with CCS facilities. In this case, carbon emission rights only have to be bought for the amount of carbon dioxide emissions released into the atmosphere. The carbon sequestration process, the transport to the storage facilities as well as their operation give rise to additional costs.

The discounting of both annualised investments and other costs, incorporated through the factor $\delta_{a\alpha p}$ plays a central role in the investment process. The central aim of the process is to integrate the perspective of the investors, namely the reinvestment premise. In the model, all costs are discounted to one arbitrary *discounting base year*, implicitly comparing their capital value. In PowerACE-Europe, the net present value method is applied in slightly modified version: usually, the initial investment in t_0 is not discounted if t_0 is the base year. In the model, the investment is first annualised and then discounted again to the discounting base year, which is counter-intuitive at first. This has to be done as the years that are not part of the calculation, (e.g. 2021, 2022, etc.) would otherwise influence and bias the ratio of fixed and variable cost included in the calculation, indirectly penalising capital intensive technologies with long lifetimes. The implications are discussed in detail section 4.5.3.

Furthermore, several additional equations are set up by the CPM in order to take national characteristics into account. Firstly, the construction of new nuclear power plants is excluded for countries which either

a) have a nuclear phase-out policy in place, or

b) have never shown any interest in utilising nuclear energy, for example due to the location or the existence of local, less problematic resources.

As of January 2012, this the case for Austria, Belgium, Cyprus, Denmark, Germany, Greece, Ireland, Italy, Luxembourg, Malta, Netherlands, Portugal, Sweden and Switzerland.

The construction of new lignite power plants is linked to the existence of significant domestic lignite reserves, since its low energy density renders its trading economically unattractive. Information about existing lignite fired plants is derived from the WEPP database, while the information about existing reserves stems from Euracoal, the European association for Coal and Lignite (Euracoal, 2011). For countries with sufficient lignite resources, the total capacity installed in lignite power plants is restricted to today's levels.[11] This is done as the mining of lignite is seen as a severe intervention into the landscape and the environment and attempts to significantly expand production seem unlikely to succeed.

4.3.3. Transmission grid manager

The *Transmission Grid Manager* (TGM) is responsible for defining the bounds describing the import and export of electricity. The underlying grid model can be characterised as a transshipment model as described in section 3.5.4. The process is started by importing geographical data into the system. The interconnections between all considered scenario regions are characterised mainly by their *virtual length*. The virtual length is defined here as the linear distance between the weighted load centres of the regions/countries, following the concept applied in ECF (2010). For smaller countries or countries with a relatively homogeneous settlement, the difference is small, but, for example, for the Nordic countries the load centres deviate significantly from their geographic centres. The virtual length is used for the deter-

---

[11]The more direct approach would be to limit the electricity generation from lignite or the fuel consumption. However, as lignite power plants remain to be base-load power plants in the model, the chosen approach has a very similar effect while using significantly less computational resources.

mination of the costs for expansion of the respective transmission capacities between the regions and the losses occurring on the respective lines, as shown in the eqn. (4.12) and (4.13).

$$\text{COST}_t^{fix} = \frac{(1 + i^{\text{tr}})^{n_{\text{tr}}} \cdot i}{(1 + i^{\text{tr}})^{n_{\text{tr}}} - 1} (I^{\text{tr,on}} \cdot L_t^{\text{on}} + I^{\text{tr,sub}} \cdot L_t^{\text{sub}}) \tag{4.12}$$

$$\varsigma_t = \frac{\varsigma^{\text{on}} L^{\text{on}} + \varsigma^{\text{sub}} L^{\text{sub}}}{L^{\text{on}} + L^{\text{sub}}} \tag{4.13}$$

with:

n_{tr}	assumed liftime of transmission lines p [years]
t	interconnection line index
COST_t^{fix}	annualised specific fixed cost of interconnection t [EUR/MW]
$I^{\text{tr,on}}$	specific investments of transmission lines [EUR/(km·MW)]
$I^{\text{tr,sub}}$	specific investments of submarine cables [EUR/(km·MW)]
L_t^{on}	length of the onshore part of line t [km]
L_t^{sub}	length of the submarine part of line t [km]
ς_t^{on}	relative losses of onshore lines [EUR/(km·MW)
ς_t^{sub}	relative losses of of submarine cables [EUR/(km·MW)]
ς_t	relative losses of interconnection t [1]

A distinction is made between onshore parts of transmission lines and submarine cable parts. The costs and line losses of onshore parts can be modelled in such a way that assumes a certain proportion are underground cables or DC connections. Parameters can set be for both all or individual connections, to take special circumstances like difficult terrain into consideration.

The connections are furthermore characterised by their *Net Transfer Capacity*(NTC). NTC is defined as the maximum transfer capacity compatible with security standards. As an approximation, it is the transfer capacity available for international trade. In reality, slightly different approaches for defining the maximum capacity are applied, but the general procedure defined by the TSOs is depicted in figure 4.3. The Total Transfer Capacity (TTC) is derived by increasing the transfer from a Base Case Exchange (BCE) gradually by $\Delta\text{Emax}^{+/-}$ until the security limit in one of the countries is reached. From the TTC a Transmission Reliability Margin (TRM) is subtracted, and the remaining capacity is defined as the NTC.

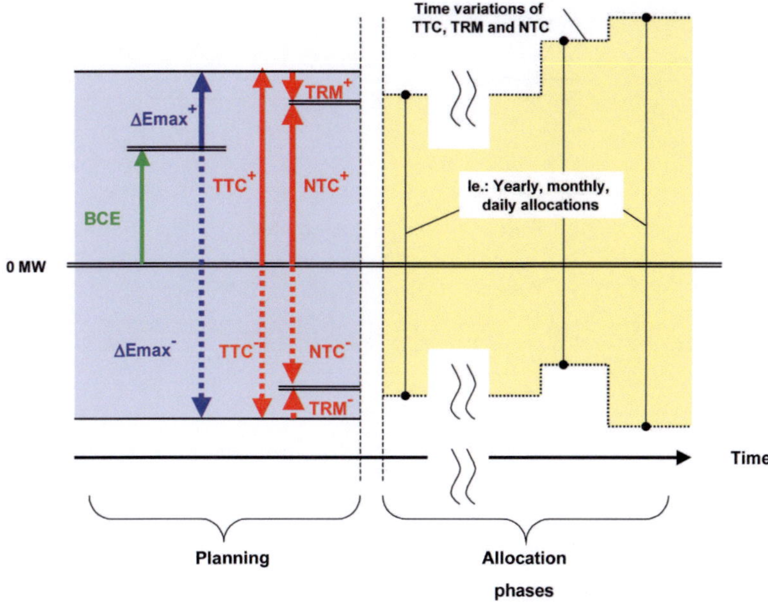

Figure 4.3.: Basic approach of defining the NTC as performed by TSOs. Abbreviations: BCE: Base Case Exchange; TTC: Total Transfer Capacity; TRM: Transmission Reliability Margin. Source: ENTSO-E (2001).

The model describes the interconnections between countries through the available NTC and the relative losses occurring when transporting electricity on the respective connection. The NTCs of the existing grid are included in the model based on information provided by ENTSO-E (2012b). The calculation of other parameters of the grid, for example the TRM is not possible at the chosen level of detail.

Finally, the TGM defines the linear equations describing the costs and bounds of the electricity transmission system as shown in eqns. (4.14) to (4.17) and transmits them to the LPM. The contribution of the transmission grid to the objective function mainly comprises fixed costs arising from the capacity expansions. However, the utilisation of the interconnection is penalised with a small term ε^t. The reason for this lies in the behaviour of the model in times of excess electricity available: in hours of highly negative residual load, resulting in curtailment of generation potential, the model may transfer electricity arbitrarily before deciding to curtail it, as the electricity has a value of zero in these hours in the model. The penalty term prevents the model from transporting power without a contribution to the objective function. Applying the term thus increases the comprehensibility of the results, as excess electricity can be clearly attributed to the region creating it.

$$\text{minimise:} \quad \ldots + \sum_{a \in \mathcal{A}} \sum_{\forall t \in \mathcal{T}} \delta_{at\alpha} X_{ta} \text{COST}_t^{fix} + \sum_{a \in \mathcal{A}} \sum_{\forall h} \sum_{\forall t \in \mathcal{T}} \varepsilon^t x_{aht}^{\text{Ex}} + \ldots \quad (4.14)$$

subject to:

$$0 \leq \qquad x_{aht}^{\text{Ex}} \qquad \leq X_{ta} \qquad \forall a, h, t \qquad (4.15)$$

$$x_{aht}^{\text{Im}} \qquad = (1 - \varsigma_t) x_{aht}^{\text{Ex}} \qquad (4.16)$$

$$[DS] \ldots + \sum_{\forall t \in \mathcal{T}_i^{\text{Im}} c} x_{aht}^{\text{Im}} - \sum_{\forall t \in \mathcal{T}_c^{\text{Ex}}} x_{aht}^{\text{Ex}} + \ldots \geq D_{ahc}^{\text{res}} \qquad \forall a, hc \qquad (4.17)$$

with:

$\mathcal{T}$ set of possible interconnection lines t

X_{ta} transmission capacity of interconnection t

$\delta_{a\alpha p}$ discounting factor for costs incurring in a for transmission t capacity built in α

ς_t relative losses occurring in connection t [%]

ε^t penalty term for transmissions [EUR/MWh]

x_{aht}^{Ex} power flow from the exporting country of line t [MWh]

x_{aht}^{Im} power flow to the importing country of line t in hour h of year a [MWh]

4.3.4. Pumped Storage Manager

The inclusion of pumped hydro electric storage (PHES) plants and electricity storage in general is challenging from a computational point of view. The description in a linear problem is highly inter-temporal: the utilisation of the facilities does not only depend on the installed capacities, but potentially on the utilisation in all other hours of the simulation. This particularity substantially increases calculation time. Therefore, the formal implementation involves a trade-off between technical accuracy and realism on the one hand, and available computational resources and runtime on the other.

A wide range of different electricity storage technologies exists, aimed at different, though sometimes overlapping, fields of application. A possible differentiation is the installed capacity in terms of rated power, ranging from small batteries in the milliwatt range to large scale applications in the megawatt range. An overview and assessment of large-scale electricity storage technologies is given by F. Genoese and Wietschel (2011).Currently only large scale applications are considered in PowerACE-Europe. As of today, these are the only ones with a significant impact on a system level; whether or when small-scale applications, like plug-in electric vehicles, become relevant on a system-level, is beyond the capabilities of the model to judge. Furthermore, storage technologies can be categorized by the size of the storage relative to the installed capacity, for example by the number of hours a facility could generate electricity and full output. As of today, the most widespread application are "day storages", as they utilise daily price-fluctuations to make profit. A typical behaviour of day storages is shown in figure 4.4 using the example of Germany: PHES are typically filled at night, when electricity prices are low and generate electricity during the daytime in times of high prices. A typical size of the storage is eight hours of full-load production.

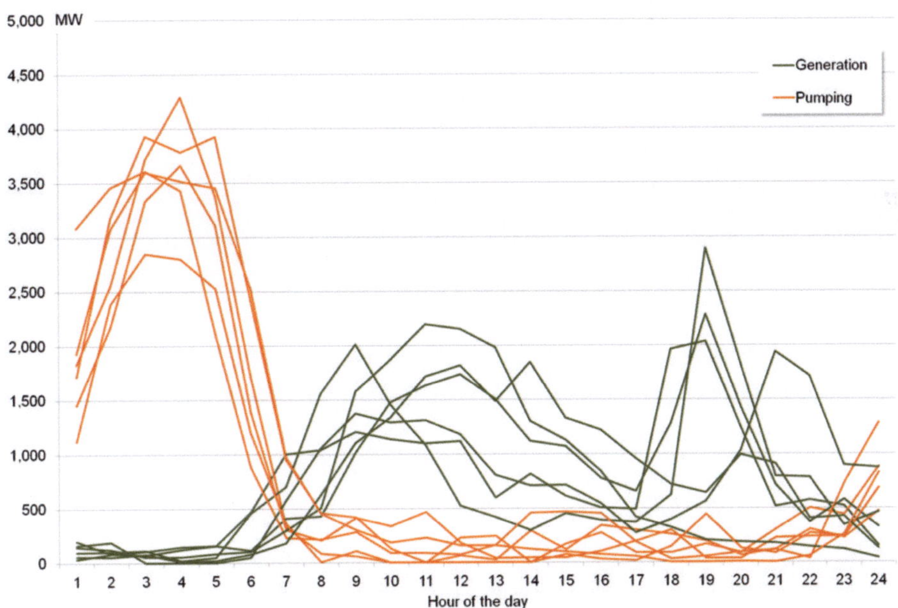

Figure 4.4.: Power generation and pumping activities of the German PHES for the third Wednesday of the months January to June 2010. Own illustration based on data of Destatis (2011).

Two types of day storage technologies are covered by the model: PHES plants and advanced adiabatic compressed air energy storage (AA-CAES) plants. PHES are the most common solution for storing large amounts of electricity and still the only commercial option to store several hundred MWh. This might change once large-scale AA-CAES become commercially available, which, according to F. Genoese and Wietschel (2011), are the second most economic options for day storage. Other storage types and applications are not considered in the model. An interesting option for storing electricity over a longer time horizon could be hydrogen. However, hydrogen becomes economically attractive only if the co-benefits, such as the utilisation as fuel, are also considered. It is not possible to adequately represent this with PowerACE-Europe, because the model is limited to the electricity sector. When modelled solely as electricity storage option, hydrogen facilities are not competitive to hydro or compressed air storages due to the low total efficiency, despite being advantageous in terms of volumetric energy density. The results of this limitation to these two technologies is discussed in section 6.3.

Similar to power plants, storage facilities are modelled in an aggregated form, meaning that all facilities able to storage electricity are modelled as one *virtual storage dummy* per country.[12] The dummies are used to approximate the optimal storage capacity without being too explicit about the applied technology. This is necessary as the data on potential sites for these two storage technologies is very scarce: to the best knowledge of the author, so far no study has comprehensively analysed their potential for Europe in a quantitative way, to be used as data source for modelling the available potential. The few existing studies are too selective, i.e. deal with one country, region or specific site and only cover one type of technology. The dummies however, represent both technologies. Once data on potential sites becomes available, it will be integrated in the model.

The virtual storage dummy approach only seems admissible under certain conditions. First of all, the technologies to be aggregated have to be comparable in their techno-economic properties. In all the following scenarios, the aggregation only covers pumped storage hydropower and AA-CAES plants. The specific investments in both technologies have to be generalised and estimated, since the investments for hydro power plants vary greatly from one location to another; the future costs of large AA-CAES are highly uncertain. Nevertheless, it seems plausible that by 2020 costs will be in the ranges depicted in figure 4.5. For both systems, a storage

---

[12]It is possible to model several storage technologies simultaneously with individual parameters, though this is not applied in the scenarios presented here.

allowing eight hours of electricity conversion at peak capacity is assumed.[13] The system efficiency is set to 80 %, which is typical for PHES plants, but will probably not be reached by AA-CAES facilities in the mid-term. Therefore, the results of the model in terms of optimal storage capacities should be seen as upper estimates. Furthermore it has to be noted, that as it will become evident from the results in chapter 6, the model concludes a very moderate expansion of storage facilities under the given parameters. Overall, the impacts of the simplification are considered to be within reason.

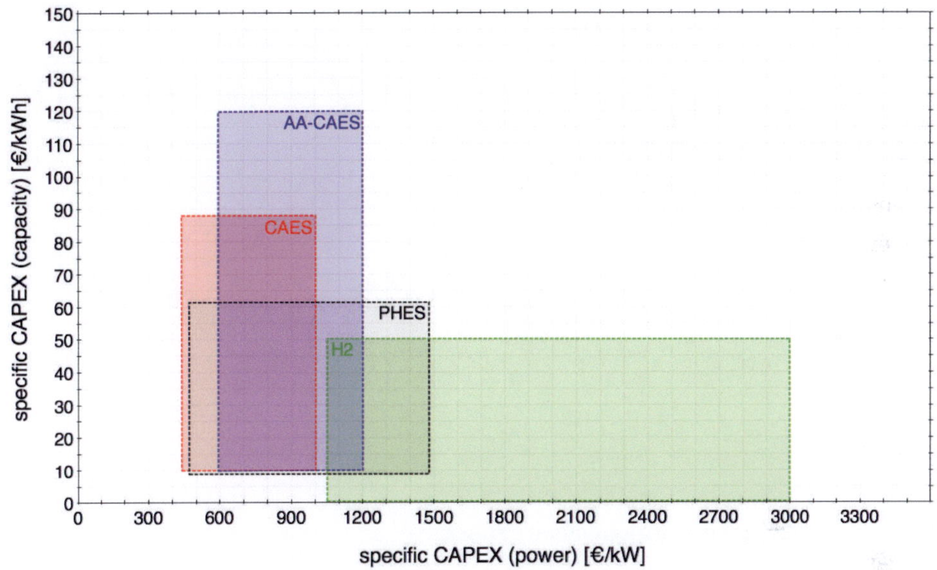

Figure 4.5.: Typical cost ranges of electricity storage facilities. Source: Internal database of Fraunhofer ISI based on published data of existing systems and cost prognoses.

The Pumped Storage Manager (PSM) loads the data on the existing and planned pumped storage hydropower plants, which is the only electricity storage technology with noteworthy current capacity. The information on the reservoir size is disregarded, while the pumping and generating capacities are totalled. The size of the storage is fixed relative to the installed capacity; for both systems, a storage of eight hours of electricity conversion at peak capacity is assumed. As the lifetime of the storage facilities is assumed to be greater than 40 years, re-investment processes can be ignored in the LP's formal description and the technical lifetimes plays a role only the calculation of the annualised costs.

---

[13]Other storage size have been been tested, with small effects on the economic attractiveness of the storage facilities. See section 6.3 for details.

$$\text{minimise:} \quad \ldots + \sum_{a \in \mathcal{A}} \sum_{\forall \alpha \in \mathcal{A}_a} \sum_{\forall c \in \mathcal{C}} \sum_{\forall s \in \mathcal{S}} \tau_{as\alpha} \delta_{as\alpha} X_{cs\alpha} \text{COST}_{s\alpha}^{fix} \tag{4.18}$$
$$\underset{\vec{X},\vec{x}}{}$$

subject to:

$$\text{CAP}_{acs}^{tot} = S_{cs}^{start} + \sum_{\forall \alpha \le a} X_{cs\alpha} \qquad \forall a, h, c, s \tag{4.19}$$

$$0 \quad \le x_{ahcs}^{turb} \le \text{CAP}_{acs}^{tot} \qquad \forall a, h, c, s \tag{4.20}$$

$$0 \quad \le x_{ahcs}^{pump} \le \beta_s \text{CAP}_{acs}^{tot} \qquad \forall a, h, c, s \tag{4.21}$$

$$0 \quad \le \text{F}_{ahcs} \le \text{R}_s \text{CAP}_{acs}^{tot} \qquad h < h^{max}, \forall a, c \tag{4.22}$$

$$\tfrac{1}{2}\text{R}_s\text{CAP}_{acs}^{tot} \le \text{F}_{ahcs} \le \text{R}_s \text{CAP}_{acs}^{tot} \qquad h = h^{max}, \forall a, c \tag{4.23}$$

$$\text{F}_{ahcs} = \begin{cases} \tfrac{1}{2}\text{R}_s\text{CAP}_{acs}^{tot} & \text{if } h = 0, \forall a, c, s \\ \\ \text{F}_{a(h-1)cs} - \sqrt{\mu_s}\, x_{a(h-1)cs}^{turbine} + \sqrt{\mu_s}\, x_{a(h-1)cs}^{pump} & \text{if } h > 0, \forall a, c, s \end{cases} \tag{4.24}$$

$$[DS]: \ldots + x_{ahcs}^{turb} - x_{ahcs}^{pump} + \ldots \ge D_{ahc}^{res} \qquad \forall a, h, c, s \tag{4.25}$$

with:

s electricity storage facility index

$\mathcal{S}$ set of available storage facility technologies s

$X_{cs\alpha}$ investment decision variable, i.e. capacity increase in storage technology s in year α in country c

$\tau_{as\alpha}$ Binary variable, "1" if $a - \alpha <$ the technology's lifetime, "0" otherwise.

$\delta_{a\alpha s}$ discounting factor for costs incurring in a for capacities of technology s built in α

x_{ahcs}^{turb} net electricity generated in year a in hour h in country c by storage technology s [MWh$_{el}$]

x_{ahcs}^{pump} net electricity withdrawn from the grid in year a in hour h in country c by storage technology s [MWh$_{el}$]

β_s Assumed ratio of pump capacity to turbine capacity [1]

F_{ahcs} Summed filling level of the storages technology s in year a , country c and hour h

While eqns. (4.19) to (4.21) set the operational limits of the pumps and turbines of the virtual storage dummies, eqns. (4.22) to (4.24) describe the filling level of the storage and its hourly limits. The storages start each year with filling level of 50 % and have to be in the same state again in the last hour.

4.3.5. Renewable Energy Manager

Power generated from renewable energies is considered in four different ways in the model, i.e. as

1. exogenous data on *non-dispatchable* RE technologies,

2. exogenous data on *dispatchable* RE technologies,

3. endogenously calculated *fluctuating* RE technologies and

4. endogenously calculated *controllable* RE technologies.

All RE technologies can generally be handled in form of exogenous input data, as this is the standard for the PowerACE model cluster.[14] For RE technologies that are modelled as exogenous and non-dispatchable, the total generation over the year is broken down into hourly values, applying the profile types indicated in table 4.3. For technologies with a rather constant profile, such as geothermal power, generation is assumed to be constant throughout the year. The same applies to technologies with a currently low relevance, for which no strong diffusion is foreseeable in the future, such as tidal power. Electricity generation from biomass, either solid, liquid and or gaseous is assumed to be partially non-dispatchable: In all scenarios presented in this work, 25 % of both installed capacity and generation throughout the year is attributed to inflexible generation, for example from heat-led combined heat and power (CHP) plants. For these, as constant profile is assumed, whereas the remaining part is treated as flexible generation from electricity-led CHP plants or plants generating solely electricity, the modelling of which is explained below. Only run-of-river hydropower plants are modelled as non-dispatchable, with monthly profiles being applied in all countries for which data is available, which is the case for most countries of the EU-27, Norway and Switzerland (EU-27+2). The exogenous non-dispatchable generation is subtracted from the total system load, thus forming the residual load D_{ahc}^{res}.

---

[14]Consequently, the loading and handling of the data is performed by methods which are part of the existing model components and not the work of the author.

Table 4.3.: Applied profiles types and data sources of exogenous RES-E. In this thesis, wind and solar power are treated endogenously.

RE technology	Profile type	Profile data source
Biogas	constant	-
Bioliquids	constant	-
Biomass	constant	-
Biowaste	constant	-
Geothermal	constant	-
Hydro	monthly profiles	ENTSO-E, Eurostat
Landfill	constant	-
Sewage	constant	-
PV	hourly profiles	model calculations based on weather data
Solar thermal	constant	-
Tidal power	constant	-
Wave	constant	-
Wind offshore	hourly profiles	model calculations based on weather data
Wind onshore	hourly profiles	model calculations based on weather data

For the RE technologies which fluctuate significantly at an hourly[15] level, detailed hourly profiles are applied. This is the case for wind power, both on and offshore as well as PV. For generating the feed-in profiles two models, *ISI-Wind-Europe* and *ISI-PV-Europe*, are applied. Both models follow a bottom-up approach and seek to calculate the aggregated feed-in profile of a certain technology in a certain region based on technological and meteorological data. It has to be noted that the calculation of the feed-in profiles is not the work the author of this thesis.[16]

For wind power, the data of weather stations of the Swiss meteorological company Meteomedia AG is processed. The positions of the 3,097 stations distributed over Europe are depicted in figure 4.6. The datasets contain information on wind speeds, local temperature and air densities. In order to calculate the power output on certain sites, the model combines the data with technical wind turbine parameters, e.g. hub-heights and exemplary power curves of existing wind turbines. The future development of the parameters is modelled in scenarios. To create an aggregated profile of all wind turbines installed in one region or country, ISI-Wind-Europe

---

[15] As already discussed in chapter 2, power output of fluctuating RE plants such as wind turbines can change considerably within one hour. However, sub-hourly fluctuations are not considered in the model since the fluctuations tend to decrease when considering generation from RE plants distributed over larger regions and due to low availability of adequate data.

[16] Both models were developed by Gerda Schubert and applied for generating the RE feed-in profiles applied in this work. The models are discussed in greater detail for example in Schubert (2012).

weights the individual profile by attributing the currently installed wind power capacities to the closest measurement station. The data is calibrated using published national wind feed-in time series in order to reach a high data fit. Naturally, a positive correlation exists between the density of weather stations per country and the accuracy of the generated feed-in profiles.

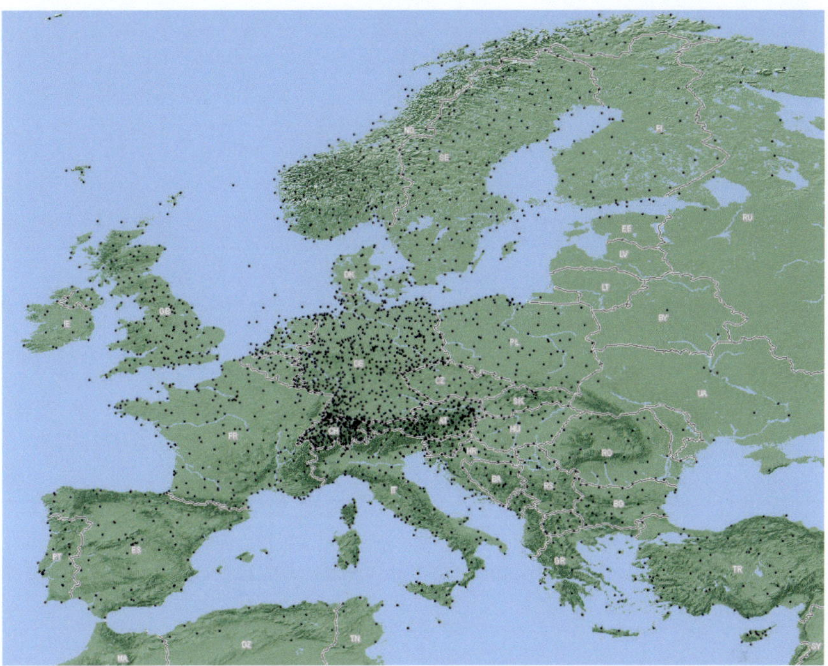

Figure 4.6.: Distribution of weather stations of the Meteomedia AG used for deriving the wind profiles by ISI-Wind-Europe.

For PV, irradiation time series are derived from satellite data, obtained from the company SoDa Service. The model *ISI-PV-Europe* processes the time series for virtual measurements stations, which are distributed at a distance of 0.5 times 0.5 degrees of longitude and latitude, as indicated in figure 4.7. Similar to the approach applied for wind energy, the model uses scenarios of the future technical characteristics of the PV modules. The electricity generation at a certain site is influenced by module and installation type, orientation and tilt angle as well as shading of the site. Furthermore, temperature data of the Meteomedia weather stations is used for approximating module temperatures. The calibration and validation process is different from the one applied in ISI-Wind-Europe, as national profiles for PV are only available for a limited number of countries and a relatively short period of time (i.e. one year or shorter).

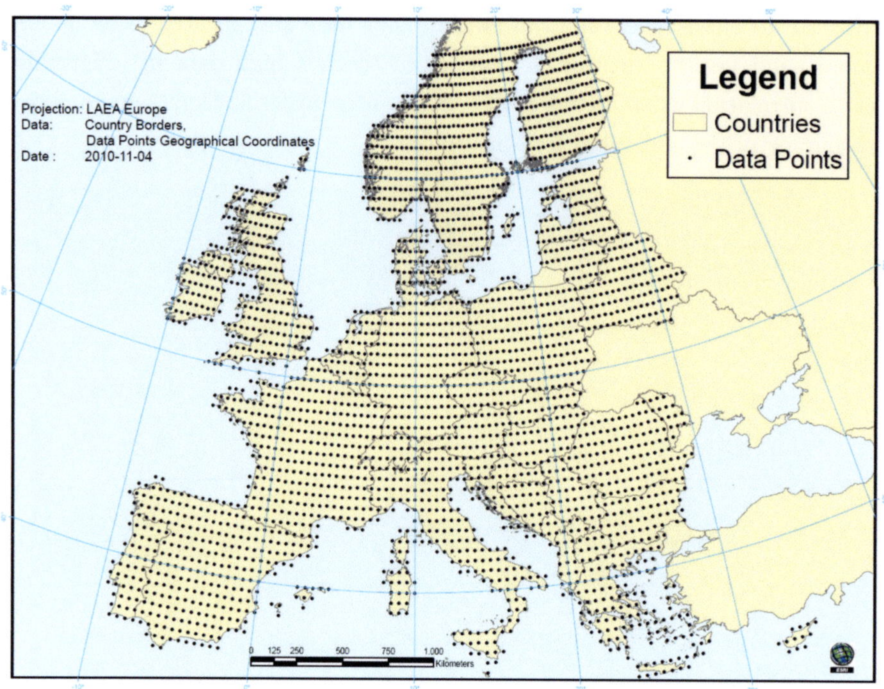

Figure 4.7.: Distribution of data points used as virtual PV measurement stations by ISI-PV-Europe.

Reservoir hydropower and a share of the generation from biomass fuels is modelled as fully or partially dispatchable. Both are treated similarly in the model: the total generation throughout the year, as defined by the scenario data, has to be met (see eqn. (4.26)), while the generation in each hour is limited only by the installed capacity (see eqn. (4.27)). The model uses scenarios containing the respective installed capacity per country and annual generation. For reservoir hydropower, the currently installed capacities and average generation in previous years are stored and applied in most scenarios without major changes. This is due to the fact that the number of potential additional sites for reservoir dams seems to be very limited and plans for their construction usually face public opposition. For biomass, it is possible to define a certain proportion of installed capacity and generation to be heat-led CHP plants, which means that constant profile is applied.[17]

---

[17]It seems plausible that in reality power generation in CHP plants is less flexible in winter due to higher heat demand than in other seasons. Furthermore, the electricity-led share of CHP

$$\sum_{\forall a} x_{ahcd} \quad = \text{GEN}_{acd} \quad \forall a, c, d[1ex] \tag{4.26}$$

$$x_{ahcd} \quad \leq \text{CAP}_{acd} \quad \forall a, h, c, d[1ex] \tag{4.27}$$

$$\text{(D-S):} \quad \ldots + x_{ahcd} \quad \geq D^{\text{res}}_{ahc} \quad \forall a, h, c, s \tag{4.28}$$

with:

d dispatchable RE technology index

x_{ahcd} net electricity generated in year a in hour h in country c by dispatchable RE technology d [MWh]

CAP_{acd} Net capacity available in year a in country c in dispatchable RE technology d [MW]

GEN_{acd} Total power generation in year a in country c from dispatchable RE technology d [MWh]

In the current model version, the optimal diffusion of four different RE technologies can be calculated endogenously. These technologies are wind power, both on and offshore, PV and CSP. While the latter is at least partially dispatchable due to the technically indispensable storage facilities,the first three are non-dispatchable. Therefore, the formal descriptions of the technologies differ.

The available potential of all RE technologies with endogenously calculated diffusion are described through cost-potential curves.[18] PowerACE-Europe aggregates and augments data originating from a GIS model. Costs are calculated on the basis of natural resources, i.e. wind speed or solar irradiation, terrain or water depth, distance to the existing grid and economic data on the respective RE technologies. For the formal description in the optimisation model the data consists of a set of *potential steps* per country, each containing information about the cost at which electricity can be generated at the underlying site. The cost data is broken down according to the respective construction years in order to take effects such as technological learning into account. Each potential step is limited in terms of capacity that can be constructed, and is measured in MW of installed capacity. Furthermore,

---

plants could also follow a daily or weekly cycle. Once data on these issues becomes available with sufficient temporal resolution it should be integrated.

[18]The components of the PowerACE responsible for calculating the cost-potential curves are not the work of the author of this thesis. The approach behind it is based on an *Geographic Information System* (GIS) and was developed by Dr. Frank Sensfuß and Dr. Martin Pudlik at the Fraunhofer ISI and is summarised in Zickfeld and Wieland (2012).

each potential step has a particular hourly feed-in profile. The number of potential steps depends on the chosen granularity. In a typical scenario covering Europe, the four technologies are described through 750 to 1,000 potential steps. For the non-dispatchable technologies (wind power and PV) the formal description is given in eqns.(4.29) to (4.32).

$$\text{minimise:} \quad \dots + \sum_{a \in \mathcal{A}} \sum_{\forall \alpha \leq a} \sum_{\forall c \in \mathcal{C}} \sum_{\forall r \in \mathcal{R}_c^{\text{flux}}} \tau_{a\alpha r} \delta_{a\alpha r} X_{r\alpha} \text{COST}_{r\alpha}^{fix} \qquad (4.29)$$

subject to:

$$\sum_{\forall \alpha \leq a} X_{\alpha r} = \text{CAP}_{ar}^{\text{tot}} \leq \text{CAP}_r^{\max} \quad \forall a, h, c, s \qquad (4.30)$$

$$0 \leq X_{\alpha r} \quad \forall a, h, c, s \qquad (4.31)$$

$$[DS]: \quad \dots + \rho_{hr}^{\text{flux}} CAP_{ar}^{\text{tot}} + \dots \geq D_{ahc}^{\text{res}} \quad \forall a, h, c \quad \forall_{r \in \mathcal{R}_c} \qquad (4.32)$$

with:

r — index of the renewable energy potential step

$\mathcal{R}_c^{\text{flux}}$ — set of all potential steps of country c with fluctuating RES[19]

$X_{\alpha r}$ — investment decision variable, i.e. capacity increase in RE technology potential step r in year α [MW]

$\tau_{a\alpha r}$ — Binary variable, "1" if $a - \alpha <$ the renewable technology's lifetime, "0" otherwise.

$\text{CAP}_{ar}^{\text{tot}}$ — total capacity installed in potential step r in year a [MW]

ρ_{hr}^{flux} — feed-in profile of potential step r in hour h [MWh/MW]

$\delta_{a\alpha r}$ — discounting factor for costs incurring in a for capacities of technology s built in α [1]

As evident from eqn.(4.32), the generation in each hour depends solely on the installed capacity $\text{CAP}_{ar}^{\text{tot}}$ and the profile ρ_{hr}^{flux} of the respective potential steps. Consequentially no hourly variables have to be determined by the model for these technologies. This also means that the process of curtailment is not explicitly performed for RE technologies. Instead, it is performed implicitly in the demand-supply equation, which is explained in section 4.3.6.

The formal representation of the dispatchable RE technology CSP requires explicit decisions about the hourly utilisation of the incoming solar resources. Eqn.

(4.35) defines that the captured thermal energy can be used either directly in the steam turbine or stored in the thermal storage. The maximum generation is limited through eqn. (4.36) by the installed peak capacity, i.e. the summed size of the turbines. The storage is subject to several restrictions, which are very similar to the ones applied for electricity storage facilities. As can be seen, the heat flows in the CSP units are described as electricity equivalents, which has to be taken into account when defining the hourly profiles.

$$\underset{\vec{X}}{\text{minimise:}} \quad \ldots + \sum_{a \in \mathcal{A}} \sum_{\forall \alpha \leq a} \sum_{\forall c \in \mathcal{C}} \sum_{\forall r \in \mathcal{R}_c^{\text{csp}}} \tau_{a\alpha r} \delta_{a\alpha r} X_{r\alpha} \text{COST}_{r\alpha}^{fix}$$

$$+ \sum_{a \in \mathcal{A}} \sum_{\forall c \in \mathcal{C}} \sum_{\forall h} \sum_{\forall r \in \mathcal{R}_c^{\text{csp}}} \varepsilon^{\text{csp}} (x_{arh}^{\text{storIn}} + x_{arh}^{\text{storOut}}) \tag{4.33}$$

subject to:

$$\sum_{\forall \alpha \leq a} X_{\alpha r} \quad = \quad \text{CAP}_{ar}^{\text{tot}} \quad \leq \text{CAP}_r^{\text{max}} \qquad \forall a, h, c, s \tag{4.34}$$

$$x_{arh}^{\text{direct}} + x_{arh}^{\text{storIn}} = \rho_{hr}^{\text{csp}} \text{CAP}_{ar}^{\text{tot}} \qquad \forall a, h, c \quad \forall_{r \in \mathcal{R}_c} \tag{4.35}$$

$$0 \quad \leq x_{arh}^{\text{direct}} + x_{arh}^{\text{storOut}} \leq \text{CAP}_{ar}^{\text{tot}} \qquad \forall a, h, c \quad \forall_{r \in \mathcal{R}_c} \tag{4.36}$$

$$0 \quad \leq \quad F_{ahr} \quad \leq R_r \text{CAP}_{ar}^{\text{tot}} \qquad h < h^{\text{max}}, \forall a, c \tag{4.37}$$

$$\tfrac{1}{2} R_s \text{CAP}_{ar}^{\text{tot}} \leq \quad F_{ahr} \quad \leq R_r \text{CAP}_{ar}^{\text{tot}} \qquad h = h^{\text{max}}, \forall a, c \tag{4.38}$$

$$F_{ahcs} \begin{cases} 0 & \text{if } h = 0, \forall a, c, r \\ F_{a(h-1)r} - \sqrt{\mu_r}\, x_{a(h-1)r}^{\text{storOut}} + \sqrt{\mu_r}\, x_{a(h-1)r}^{\text{storIn}} & \text{if } h > 0, \forall a, c, r \end{cases} \tag{4.39}$$

$$[DS]: \ldots + x_{arh}^{\text{direct}} + x_{arh}^{\text{storOut}} + \ldots \geq D_{ahc}^{\text{res}} \qquad \forall a, h, c \forall r \in \mathcal{R}_c^{\text{csp}} \tag{4.40}$$

$$X_{\alpha r}, x_{arh}^{\text{direct}}, x_{arh}^{\text{storIn}}, x_{arh}^{\text{storOut}} \geq 0 \qquad \forall a, h, c, s \tag{4.41}$$

with:

$\mathcal{R}_c^{\text{csp}}$ set of all CSP potential steps of country c

ε^{csp} penalty term for the utilisation of the thermal storage [EUR/MWh]

x_{arh}^{storIn} Electric equivalent of the thermal inflow into the storage in hour h at potential step r [MWh]

$x_{arh}^{\mathrm{storOut}}$	Electricity generated in the CSP plants' turbines in hour h at potential step r with heat from the plants' storage [MWh]
$x_{arh}^{\mathrm{direct}}$	Electricity generated in the CSP plants' turbines in hour h at potential step r with heat directly from the collectors [MWh]
ρ_{hr}^{csp}	Electric equivalent of the solar energy captured per MW installed capacity by the collectors at potential step r in hour h[MWh/MW]
F_{ahcr}	Aggregated electricity stored in potential step r in year a, country c and hour h

4.3.6. Demand manager

The *Demand Manager* administers the hourly demand-supply equations during the set-up of the optimisation problem. The full formulation is shown in eqn. (4.42). All variables contained in it have been introduced in the previous section.

$$
x_{ahc\hat{p}} + x_{ahcp\alpha} + \sum_{\forall t \in \mathcal{T}_c^{\mathrm{Im}}} x_{aht}^{\mathrm{Im}} - \sum_{\forall t \in \mathcal{T}_c^{\mathrm{Ex}}} x_{aht}^{\mathrm{Ex}} + x_{ahcs}^{\mathrm{turb}} - x_{ahcs}^{\mathrm{pump}} + x_{ahcd}
$$

$$
+ \rho_{hr}^{\mathrm{flux}} CAP_{ar}^{\mathrm{tot}} + x_{arh}^{\mathrm{direct}} + x_{arh}^{\mathrm{storOut}} \geq D_{ahc}^{\mathrm{res}} \qquad \forall\, a, h, c, \hat{p}, p, d, r \qquad (4.42)
$$

As already explained, the right-hand side of the inequation represents the residual load, which is defined in this context as the hourly demand less all exogenous, non-dispatchable generation. The left-hand side contains electricity generation as well as additional demand from storage facilities or export to neighbouring regions. The bound is formulated as an inequation. In reality, any significant mismatch between supply and demand endangers the stability of the system. The reason for the chosen approach is that it does not require a merit-order of curtailment to be calculated by the model. In reality, if generation from RES exceeds what can be absorbed and "consumed" by the grid, some generator units will be curtailed. In the current formulation this would happen arbitrarily since the variable costs of all RE technologies, except for biofuel technologies, are assumed to be zero. Therefore, the slack of the bound, i.e. the difference between the left- and right-hand side of the inequation are defined as power to be curtailed. Since in reality, little experience exists on the order in which RE plants should be curtailed, in the following it is assumed that all RE technologies without fuel costs are cut proportionally.

The LPM can add several further restrictions to the linear problem, depending on the scenario's set-up. In many cases, the total emissions from the power have to be limited. In this case, eqn. (4.43) is added to the LP.

$$\sum_{\forall h} \sum_{\forall c \epsilon \mathcal{C}} \left(\sum_{\forall \hat{p} \epsilon \mathcal{P}_c} \kappa_{f_{\hat{p}}} x_{ahc\hat{p}} + \sum_{\forall \alpha \epsilon \mathcal{A}_a} \sum_{\forall p \epsilon \mathcal{P}} (1 - \mathrm{CR}_{\alpha p}) \kappa_{f_p} x_{ahcp\alpha} \right) \leq \mathrm{EM}_a^{\max} \quad \forall\, a \quad (4.43)$$

with:

$\mathrm{EM}_a^{\max}$ maximum emissions from the power sector in year a

As for all constraints, it is possible to query the dual value or Lagrange multiplier of the constraint in the optimum. Under certain circumstances, the shadow price of eqn. (4.43) can be interpreted as an approximation the carbon price necessary to reach a certain level of emissions.[20]

4.4. Solving of the linear problem with CPLEX

The full LP is then handed over to the solver software, which determines the least-cost solution meeting all boundaries. The *ILOG CPLEX Optimization Studio* by IBM, usually referred to as CPLEX, is an optimiser implemented in the programming language *C*. PowerACE-Europe communicates with it through CPLEX's Java interface. The software was chosen after initial tests with the free solver *GLPK*[21] for its significantly shorter computation time when solving large LPs.

CPLEX offers a wide range of algorithms for optimisation problems. For linear problems, the primal and dual simplex algorithm and a barrier interior point method are the best options. Both algorithms have been tested for solving the LP described above for different scenarios. CPLEX's barrier interior point method, named *Barrier*, has been highly superior in terms of calculation time in all tests: calculating a relatively small test problem with a reduced number of countries for one year , took 118 minutes applying the Barrier algorithm. Solving the same problem with a dual Simplex algorithm was aborted after 3 days. From extrapolating the values of the primal and dual objective function reached at that time it can be expected that it would have taken at least 15 days to finish the Simplex algorithm.

---

[20]This is discussed in detail in section 6.1.8.

[21]For the solver and its documentation, please refer to http://www.gnu.org/software/glpk/.

As CPLEX is a proprietary software, the exact workings of the Barrier algorithm are undisclosed. The documentation[22] is not very explicit about anything not directly related to the operation of the program. The Barrier optimiser is a primal-dual logarithmic barrier algorithm, which belongs to the group of interior point methods. The history and function of these algorithms is summarised, for example, in Wright (2004). While the steps in the Simplex algorithm proceed along the edges of the feasible solution space, interior point methods are "pulled" though its inner part through logarithmic damage functions. The objective function of any LP in standard form may be altered to

$$\min_{x} \quad c^T x - \mu \sum_{i=1}^{n} \ln x_i \tag{4.44}$$

with $\mu > 0$ often being referred to as the *barrier parameter* (Colombo, 2007). For $\mu \to 0$ the iterations converge to the optimal solution. The Barrier algorithm finishes once the relative complementarity of the problem is near zero, with a tolerance that can be defined by the user. This also means that without alteration of the results, the Barrier algorithm cannot deliver an optimal solution, but only a solution that is very close to an optimum. However, CPLEX can perform a basis crossover, thus generating the solution on the edges of the LP closest to the solution of the Barrier solution. This is usually not performed by PowerACE-Europe, as the majority of the inaccurate values are smoothed out by rounding, e.g. an installed capacity of 0.1 MW in a certain technology will appear as zero.

The Barrier algorithm, or interior point methods in general, tend to be advantageous for large, sparse problems. Sparse in this context means that although the matrix A of the LP may be large, the coefficients in most rows are zero. For these problems techniques for factorising matrices based on Cholesky decomposition are very efficient and fast. The Barrier algorithms can be parallelised, allowing the model to use several CPUs at once. The number of CPUs can be defined by the user, with the optimal number depending on the size and design of the problem. Calculation time does not strictly decrease with higher numbers of CPUs and may even increase in some cases. The calculation time also depends on other parameters, which will not be explained here in detail. In some case a trade-off has to be made between calculation time and the likelihood of the solver finding a feasible solution. Due to the large size of the problem, certain settings can result in nu-

---

[22]An introduction to CPLEX is given at www.ibm.com/software/integration/optimization/ cplex-optimiser/. The calculations for this thesis are performed using CPLEX 12.4.

merical instabilities of the algorithms, which lead to a termination of the model. The Barrier algorithm can be run in an indeterministic mode, which means that the next attempt with the same LP may be successfully solved.

After the successful optimisation of the LP the results are queried from the solver, processed and stored. The results are written both to the hard drive of the server running the model and to a MySQL database. Although the process will not be explained here as it is of no scientific interest, it should not be ignored that handling the huge amounts of data generated by the model is challenging. A single run of a typical scenario generates over a hundred million figures. The way these are structured essentially defines how quickly the model's user is able to understand and interpret the result.

4.5. Critical reflection and conclusion

The model PowerACE-Europe belongs to the group of system optimisation models, thus sharing many of the typical strengths and weaknesses of the model group. Some peculiarities and disadvantages of optimisation models in general and of PowerACE-Europe in particular have to be borne in mind when interpreting the model's results.

4.5.1. Market representation in the model

Like all electricity system models, PowerACE-Europe assumes perfect competition. Consequently the impact of market power and other distortions cannot be incorporated. However as the nature of the changes in power market regulation, and shifts in future market power would be speculative, this is not seen as a significant flaw of the model. Furthermore, as described in chapter 2, several of the regulatory changes of recent years have the explicit goal of bringing the European electricity market closer to a perfect market. Assuming such efforts of policy maker and regulatory authorities are successful, it could be assumed that the differences between a model presuming a perfect market and the real market would decrease over time.

More problematic in this context is that the results of changes in regulation cannot be analysed by the model; it might be of interest to simulate the results of certain market designs on the long-term outcome, but this is not possible with the model. Combining PowerACE-Europe with other models with a stronger focus on markets, e.g. the agent-based version of the model, can bring additional insights in this respect.

Critics to optimisation approaches in general point out that the results depict a future that in reality cannot be reached. Market distortions and imperfect knowledge ignored in the model, significantly alter the decisions of market participants, shifting reality away from a least-cost solution. Although this argument is valid, it is only a failure if one attempts to use the model to generate a forecast; the model is not suited to such a task.[23] The model delivers the least-cost strategy to reach certain targets under certain conditions, which must not be seen as forecast. The results are relevant *especially* for their deviations from reality. As an example, the model concludes that in a certain scenario only few new storage facilities should be built. This is the case for most scenarios discussed in this work but does not mean that in reality no electricity storage will be built or that the construction of new storage is not cost-efficient. The results of the model can only conclude that, under the given assumptions and having taken into account certain aspects, the analysed storage technologies are not part of the least-cost solution. This might lead to further conclusions or the identification of the need for additional research. In this case, for example, that might include an analysis of the benefits that small-scale storage could bring to lower voltage levels.

4.5.2. Perfect foresight

PowerACE-Europe applies a perfect-foresight approach. This means that investment decisions concluded for 2020 are subject to the knowledge of all relevant parameters over the subsequent decades. Naturally, this is an unrealistic assumption. Still, the current formulation and available calculation capacity would not allow for a stochastic or statistical version of the model. Optimising the decisions along one deterministic path currently takes available computational resources close to the edge; applying stochastic or statistical modelling would most likely multiply the system requirements. If this became feasible in the future, it would substantially enlarge the model's fields of application. Currently, the results should be tested with carefully selected sensitivity analysis in order to prevent results based solely on artifacts or reactions to certain parameter settings.

Nevertheless, the discounting approach, which is discussed in the following section, mitigates the impact of the perfect foresight to some extent: the knowledge of the distant future influences the decision process, but it has, depending on the settings, significantly less influence than the present or the immediate future.

---

[23]Considering the lack of success of basically all attempts to precisely forecast technological developments over several decades, one might generally question the idea.

4.5.3. The implications of the applied discounting approach

Discounting is applied in economic models mostly for one or more of the following reasons:

- *Time preference*, as individuals tend to value immediate consumption over consumption in the future. Over long time horizons, this is also linked to intergenerational fairness.

- *Time value of money*, taking into account the respective interest rates and costs of liquidity.

- *Inclusion of risk*, to take effects such as risk aversion into account, which is discussed for example in Ehrenmann and Smeers (2011).

The question of how appropriate discounting is in energy system models, or in climate policy analysis, is still debated. The large range of different approaches to the issue in academic literature is surprising and a brief overview of the positions is given by Tóth (2000). In many cases, the discussion focuses on the impact of discount rates on the analysis of strategies against long-term climate change, as discussed for example in Stern (2007). The discussion often contains ethical elements, as discounting affects the valuation of the benefits for future generations versus costs to be borne short- and medium-term.

Several key issues are relevant for the model at hand here. Firstly, the capital value approach, that is implicitly or explicitly the basis for many economic comparisons in models, is not undisputed; especially the reinvestment premise is often criticised: As Franke and Hax (2009, p. 166) point out, the method simplifies the real financing options of firms and does not take taxation into account. For example, a firm with weighted average cost of capital of 12 % per year might not be able to yield the same profit for alternative investments. This is not reflected in any approach relying on a single internal rate of return.

Secondly even low discount rates tend to marginalise the importance of events in the distant future. Figure 4.8 shows the values of the function $\frac{1}{(1+i)^n}$ for different interest rates and time intervals. Depending on the interest rate, the value after discounting to 2010 is between 9.7 and 1.1 % of the real value in 2050. In this rather techno-economic model, this leads to an effect that can be described as "procrastination": where the emissions have to be decreased below a certain value by 2050, the model can recommend late action. In other words, faced with the low weight of costs in later years, the most cost-efficient solution can be to initially build and utilise emission-intensive plants and invest in clean technologies in the final period from 2041 to 2050.

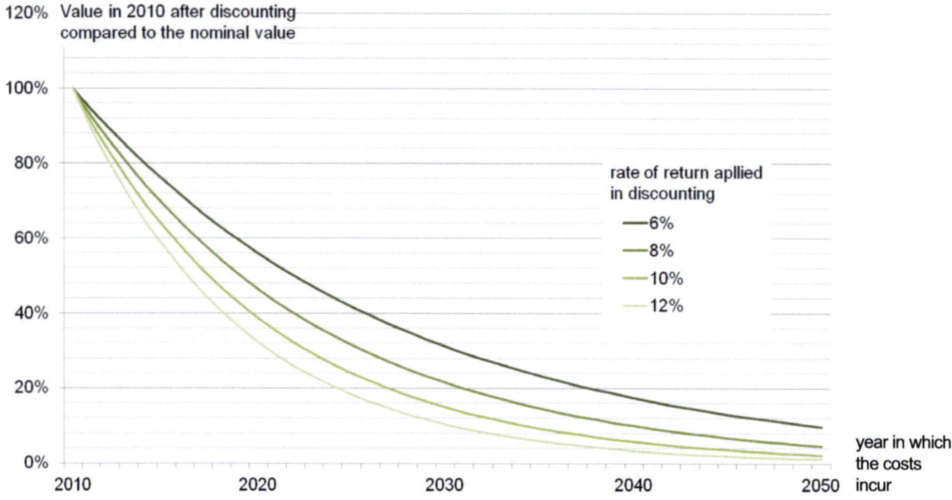

Figure 4.8.: Decrease in value when discounting costs to 2010 for exemplary interest rates.

To tackle this issue, several authors argue for the use of lower discount rates: Newell and Pizer (2004) point out that the uncertainty of future rates demands the application of values at the lower end of the possible range. Howarth (2006) argues for the use of very low discount rates below 1 %, since the risk premium could be set to negative values for policies that reduce risks to future economic welfare.

The discounting process "dilemma" is slightly different for PowerACE-Europe, or in energy system optimisation models in general; as there is no future benefit to be discounted, the model only looks for the least-cost option to fulfil a certain goal. Leaving aside questions of intergenerational fairness, the problem is thus reduced mainly to deriving an appropriate time value of money and possibly to including risks.

A discounting process as applied in PowerACE-Europe generally affects the decision by giving more weight to the earlier years of the calculation whenever positive discount rates are applied. This is exemplified in table 4.4: in this example, two hypothetical, but not unrealistic power plant investment options are shown. Both will run 5,000 hours per year and yield the same earnings. A simple addition of annualised capital costs and variable costs for the years 2010, which is the investment year, and 2030 would conclude that the combined-cycle gas turbine (CCGT) power plant can supply the electricity at lower total costs, as the variable costs of coal increase sharply in 2030. However, at a discount rate of 10 %, the coal power plant is the preferred option. In theory, the higher earnings in the earlier period could be reinvested and overcompensate the higher costs in the later period. This is a typical problem for investments in clean technologies in the electricity sector: existing incentives are often low, e.g. from the EU ETS, and can only be partially compensated for by the expectancy of higher prices in the future. Incentives to invest into low carbon technologies are dampened by the fact that the immediate future is economically more relevant.Excluding this effect, by applying zero or very low discount rates, results in the model choosing technologies that in reality are not lucrative for investors.

Table 4.4.: Example on the difference between undiscounted costs and capital value.

	Coal power plant	CCGT power plant
Interest rate	10 %	10 %
Specific investment [EUR/MW]	1,300	800
Life time [years]	40	40
Utilisation [hours per year]	5,000	5,000
Variable costs 2010 [EUR/MW]	50	65
Variable costs 2030 [EUR/MW]	90	75
Summed costs (2010 and 2030)	965,874	863,615
Capital value	469,587	474,709

It has to be noted and criticised that in the documentation of most power market or system models and in most publications building on the calculations of these models, this topic is neglected. This could be due to the focus being typically set on the technical aspects, effectively ignoring the immense impact of the issue. Nevertheless, some authors have pointed out that other reasons might exist for concealing the details of the discounting approach:

> "All the economic parameters, such as discount rate, rate of return and other project specific factors, are hard to generalise to make a fair comparison for all technologies. The results are often very sensitive to changes in these economic assumptions. It is possible to manipulate the results in a desired direction by skewing the economic assumptions: if this is done skilfully, almost every result can be produced, even within a realistic set of parameters." (PwC, 2010)

4.5.4. Further options for improvement of the model

For further improvements for PowerACE-Europe, the already discussed limitation of the model have to be kept in mind. However, several possibilities have already been identified in this chapter.

A logical next step is the inclusion of supply side activities. In a first step, the shape of the load profile has to evolve over time in order to take into account the influence of, for example, electric vehicles or heat pumps. Afterwards, demand should be able to respond to supply side situations such as over- or under-supply from fluctuating RES. Demand response could lead to lower load or shift load away from the critical hours of the system. The first option could be realised in the model as a cost-potential curve of loads that receive payments for not consuming. The second option could be implemented similar to short-term storages, allowing to delay or bring forward consumption. Electric vehicles could be integrated likewise; car users could either delay charging or allow the usage of the car batteries in vehicle-to-grid applications. It seems plausible that the inclusion of demand response has a significant impact on the results; presumably, it would reduce the need for gas turbines, storages and transmission grids.

Another area for further development is the representation of biomass in the model. The expansion of the technology should be made endogenous, which would, for example, allow research on the optimal diffusion under different fuel price scenarios. The technology could also be a point of contact for an inclusion of the heating sector.

In order to expand the possible fields of application for the model, its integration and interaction with the other parts of the PowerACE model cluster should be advanced. For example, if the solution of optimisation in terms of installed capacities was fed back into the agent-based part of the model, the performance of the power plant park under certain market rules could be bested. It would open the model to questions of market design, such as the potential necessity of a remuneration for providing capacity versus energy-only market forms.

In the medium-term future, a model expansion moving away from deterministic to stochastic optimisation seems to be necessary, despite being very ambitious. Especially when researching on robust strategies for reaching climate targets, the inclusion of different future developments in another form then scenarios seems very beneficial. This could allow for the determination of an infrastructure portfolio taking into account the future uncertainties rather then being designed to be optimal under very specific circumstances. The same accounts for wind and solar power; the results for these could be made more robust through stochastic optimisation. However, this is not only a question of methodology but also of available computational resources.

4.5.5. Conclusion

This chapter describes the creation of the model PowerACE-Europe as an extensive modification of the PowerACE simulation model. All requirements regarding the model's capabilities defined in the previous chapter could be met, although only a few of the subsidiary features could be incorporated. The model's central strengths are the very high temporal resolution and coverage. Few power system models exist which take into account both capacity expansion and unit commitment, and consider all hours of the year. To the knowledge of the author, all models that allow such a high temporal resolution (see for example: Nicolosi (2012)), cover only a single country, whereas PowerACE-Europe can handle over 50 countries. Furthermore, the model incorporates a high level of detail in its representation of renewable energies, including weather profiles based on actual meteorological data and a high spatial granularity. The combination of these features are unique for an electricity model and allow analyses that are not possible with other models: As it will be shown in chapter 6 a cost-efficient European power system with a high share of fluctuating renewable energies has to rely on inter-regional weather effects for balancing out fluctuations in the utilised RES. The high temporal coverage of PowerACE-Europe allows in-depth examination of the requirements for such a system in terms of power plants capacities, interconnections and storage facilities. The high temporal coverage of PowerACE-Europe increases the results' robustness, as the system is stable in a very high number of states and meteorological circumstances and not only in a few "typical" states. The drawback is that the approach generates an excessive amount of data, requiring a powerful server to handle it. The amount of data also limits the model's ability to depict market behaviour, which is assessed to be only a minor flaw for carrying out long-term scenarios.

In the next two chapters, a series of four scenarios will be defined and evaluated for approaching the research question at hand.

5. Definition of long-term scenarios and exogenous parameters

To approach the research questions of this work, a series of four decarbonisation scenarios is analysed using the model PowerACE-Europe. The rationale behind these scenarios is two-sided. On the one hand they can be seen as strategies for decreasing the electricity related emissions to levels in line with the 2° C target.[1] On the other hand the differences between the scenarios can also be considered as exogenous conditions.

Before the scenarios are discussed, the conscious omission of a baseline or reference scenario should be noted. This is unusual for scenario analysis of climate change mitigation. Typically, such scenarios assume either no mitigation efforts or extrapolate current policies, e.g. the current level of mitigation efforts. Thus the central motivation behind the scenarios is to provide a baseline for cost calculations. In this thesis however, no baseline/reference scenario is calculated for the following two reasons.

Firstly, since the applied model only covers the electricity sector, the differences between the baseline and the decarbonisation scenarios would largely depend on the exogenous scenario assumptions. For these, the is a large uncertainty range of justifiable values. For example, the EU ETS is currently agreed upon for an indefinite time; whether or not the it will exist up until 2050 seems to be connected to some degree to the mitigation efforts in the rest of the world. Closely related to this issue, fossil fuel prices can be assumed to be higher in scenarios with low or no CO_2 prices, but are highly influenced by global developments. In conclusion, a baseline scenario would deviate so substantially from a decarbonisation scenario in its exogenous assumptions that the endogenous cause-effect relationships would not be clearly distinguishable.

Secondly, fighting climate change has become a major goal of European energy policy. The question of whether climate change should be fought is discussed in

---

[1] "Strategy" in this context refers to the design of the power system and its decarbonisation, i.e. the complete infrastructure and its utilisation. Furthermore, to a limited degree it also refers to energy efficiency policies; however this is only included exogenously through the demand input data. The explanatory value of the results regarding demand side policies is very limited.

various publications, the most extensive so far probably being the report of Working Group I of the (IPCC, 2007b).[2] Questions of this dimension are beyond the capabilities of the model PowerACE-Europe. As an electricity system model it focuses on the direct costs of mitigation measures. Potential benefits of the applied measures, such as avoided adaptation costs and employment effects, as well as moral implications linked to insufficient action are omitted or neglected.

In consequence, this thesis explicitly only analyses *how* the power sector can contribute to fighting climate change. Nevertheless, a sensitivity analysis on costs associated with certain levels of decarbonisation of the power sector is performed in section 6.3. If similar data is available for other sectors, this approach can give an indication of the cost-efficient distribution of decarbonisation efforts among the sectors.

5.1. Understanding of the term "decarbonisation scenario"

It is impossible to precisely define decarbonisation scenarios solely for the European power sector. In the literature, the term[3] is often used to describe developments that result in a power sector complying with 450 ppm scenarios, for which the IPCC estimates a likelihood at least 50 % for reaching the 2°target (Meinshausen, 2005). The European power sector is, however, responsible for only a fraction of the worldwide total GHG emissions; in 2009, approximately 2.9 % of global emissions were caused by power generation and district heating in the EU27+2[4], corresponding to less than 5 % of worldwide emissions from fuel combustion (IEA, 2011a). Decreasing worldwide emissions will involve trade-offs between countries and sectors, which, besides aspects of polluter liability and unequal willingness or ability to pay, is also a question of determining the cost-efficient exploitation of GHG saving options: In some sectors or regions decreasing emissions is possible at significantly lower costs than in others. In recent years, a strong agreement seems to have emerged that the power sector offers comparatively cheap options for decreasing emissions. Calculations of the European Commission conclude that for the European Union to achieve an "overall reduction of 80 % [...] means that the energy sector has to bring its emissions to almost zero", as stated by (Oettinger, 2011) in the press conference for the publication of the *"Energy Roadmap 2050"*. The impact assessment pub-

---

[2] The answer to this question from a world perspective seems to be a clear "yes" due to the high costs of adaptation measure otherwise necessary in the long-term future .

[3] Other expressions, e.g. 2° C scenario or 450 ppm scenario, referring to the atmospheric CO_2 concentration, are used in a similar or identical way.

[4] Own calculations based on data from IEA (2011b) and UNFCC (2011).

lished with the roadmap reflects this conviction; in all decarbonisation scenarios presented, emissions from the power sector decrease below 52.5 Mt by 2050. This is equivalent to a reduction of 96.4 % compared to 1990 levels.

The decarbonisation scenarios that are discussed in the following sections reach a reduction by 95 % compared to 1990 levels for the EU27+2. This corresponds to annual emissions from the power sector below 75 Mt/a until 2050 in all scenarios. As previously mentioned, the impact of higher and lower reductions are tested.

5.2. Options for decarbonising the power sector

On the supply side of the power sector three technology groups can be seen as the main mitigation options. In all decarbonisation scenarios, both in this thesis and in other publications, the majority of the decrease in emissions on the supply side is reached through a combination of renewable energies, nuclear power and CCS technologies, as indicated in figure 5.1.[5] In reality, not all combinations of these options are technically feasible or desirable. Active phase-out policies and technical limitations make an electricity system with shares of nuclear power above 70 % highly unlikely, if not impossible. In fact, virtually all decarbonisation scenarios published in recent years foresee or propose RE shares above 50 % (cf. Fischedick et al. (2012)), i.e. a combination of measures in the upper half of the triangle in figure 5.1. In some cases, most prominently in the the scenario developed in Teske, Muth, et al. (2012), the decarbonisation of the supply side is achieved solely through the diffusion of RE technologies.

The options on the supply side can be complemented with measures on the demand side. Most notably, energy efficiency and conservation measures may decrease the demand for power, although the chances of success of such measures are not undisputed (see, for example, Herring (2006)). Furthermore, some demand side measures, such as *Demand Response* (DR) or *Demand Side Management* (DSM) can facilitate supply-side decarbonisation measures, often due to their ability to react to the fluctuations of RES-E. Although the general effects of reduced demand are explored in this thesis in a scenario, the complex interactions between demand and supply side cannot be completely covered.

---

[5]Other options, e.g. fuel switching, might play a supplementary role, but in the long run their possible contribution is clearly limited. Reductions in demand through energy efficiency and conservation measures usually also play a central role.

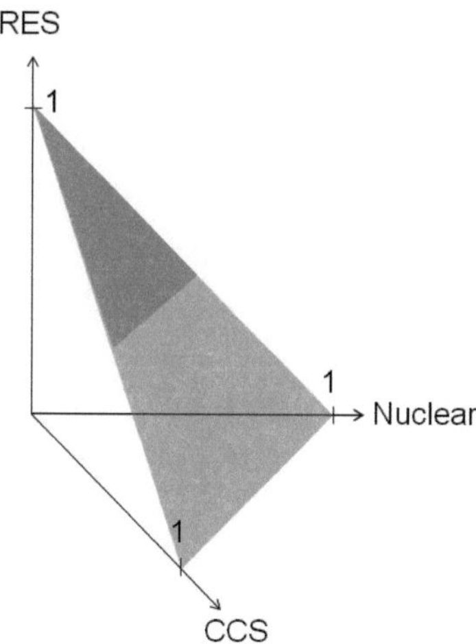

Figure 5.1.: Possible combinations of the contributions from the three major decarbonisation technologies on the supply side.

5.3. Definition of the scenarios to be analysed

To analyse the interdependencies between the different components of the infrastructure of the supply side, four scenarios are analysed. The scenarios are to some
degree designed to explore strategies and uncertainties for the future supply mix.
Major questions in this context are:

1. Will CCS technologies be available?

2. Will the development of the grid be fast enough for the transition necessary
   for the decarbonisation of the power sector?

3. Will efficiency measures be able to significantly decrease electricity demand?

However, in reality these uncertainties are not the results of coincidences, but
the results of actions of stakeholders. Therefore, from the viewpoint of European
decision makers, these uncertainties could also be phrased as questions regarding
strategy options:

1. Should CCS be pursued as a central measure in the power sector?

2. How important is an extensive and fast expansion of the interconnections
   between countries?

3. What can be gained from the successful implementation of efficiency measures?

Although there are many other uncertainties, the focus is set on these three in the following sections for their high impact on most parts of the electricity system. Table 5.1 shows an overview of the general differences in the input data of these scenarios, while all other aspects are kept constant. The differences will be summarised in the following before the input parameters and the sources are stated.

Table 5.1.: Overview over the main differences between the scenarios analysed in this work.

	Optimistic Decarbonisation	No CCS	Hampered Grid	Strengthened Efficiency
Abbreviation	OPT	NoCCS	GRID	EFF
Electricity demand	high	high	high	low
CCS technology available	yes	no	yes	yes
Prompt grid extensions	yes	yes	no	yes

5.3.1. Optimistic decarbonisation (OPT)

This scenario represents optimistic assumptions for technical and political developments regarding the supply side: all power generation technologies are available, including CCS and nuclear power. While the latter is still somewhat restricted through national phase-out policies, CCS plants can be built without restriction, implying that all technical problems can be solved and current public and political resistance will be overcome. The costs for both CCS and RE technologies decrease significantly over time. For the exogenous RE technologies, strong support policies are assumed in the model *PowerACE-ResInvest*, which is applied to calculate the diffusion paths. Furthermore, the expansion of the electricity grid is not hindered by limitations and the construction of electricity storage facilities is possible in all countries at prices similar to pumped storage hydropower.

However, the positive developments on the supply side meet challenging conditions from the demand side: The electricity demand increases steadily due to economic growth and fuel switching; many applications change need to use electricity as fuel in order to achieve the ambitious decrease in emissions. This is the case for example for private and public transport as well as for heating, where the role of heat pumps increases. Energy efficiency measures, though being implemented, are not able to adequately compensate for these increases.

The scenario represents a situation in which all of the major mitigation technologies on the supply side are available, along with the enabling and complementary technologies such as transmission grids and storages. In such a case, the perfect market assumed in the model can freely determine the cost-efficient combination and utilisation of the options. The OPT scenario acts as a point of reference for the other scenarios, in which specific assumptions are altered, allowing for the individual examination of the consequences.

5.3.2. No CCS (NoCCS)

The second scenario is identical to the OPT scenario with the exception that CCS technologies do not become available on the market. The causes of this setting are not specified: it could, for example, be a decision made by national governments due to strong public opposition or by market participants unconvinced that the existing issues can be overcome in time. The scenario does not seem unlikely (and is hence relevant to explore) due to the serious issues that the CCS diffusion currently faces (see, for example, von Hirschhausen et al. (2012)). This situation is contrasted by the fact that CCS represents a significant pillar of the EU decarbonisation policy: For example, in the Energy Roadmap and the accompanying Impact Assessment (see: European Commission (2011c,d)), only one of five scenarios decarbonises the power sector without using CCS, while in the other scenarios CCS technologies have a share of 19 to 32 % in total electricity generation.

The model has to compensate for the missing option to generate dispatchable power with low emissions. It has to be pointed out that it is assumed that the decision to refrain from CCS in the power sector takes place in the near future. Therefore, alternative paths can be chosen from the beginning on. A later decision would be a different scenario. This might be, for example, if in 2025, i.e after investments into new, theoretically "CCS-ready" power plants have been made, which then cannot be equipped with carbon sequestration.

5.3.3. Hampered Grid (GRID)

Comparable to the NoCCS scenario, the GRID scenario explores how real-life issues, if impossible to overcome, affect a least-cost solution delivered by an electricity system model. The ongoing diffusion of RE in the power sector has already revealed that reinforcing the grid is of critical importance.[6]. Over 80 % of the grid bottlenecks identified by the TSOs as relevant in 2020 are directly or indirectly the result of RES integration (ENTSO-E, 2012c). It is important that grid expansions are completed on a timely basis. However, most planned grid expansions are in fact behind time schedule due to several barriers, such as local opposition. These barriers lead to average construction times of new lines from 3 to 10 years (Olmos et al., 2011). However, especially interconnection lines between countries can face significantly graver difficulties. This can be seen on the example of the "Steiermarkleitung", a power line of approximately 100 km length connecting Austria and Switzerland. The line took 22 years from the first administrative steps until construction was finished in 2009 (Verbund AG, 2009).

The grid scenario aims to analyse the impacts of grid expansions being significantly below the levels calculated to be cost-efficient in the OPT scenario. Therefore, the increase in grid strength, measured in NTC multiplied by the length of the interconnection, is limited to ten years behind the OPT scenario. This approach does not capture all issues resulting from a suboptimal pace of grid expansions, because the model "knows" the issues beforehand, due to its perfect foresight, and can compensate for them accordingly. This could result in several changes, for example the construction of additional storage facilities or higher curtailment of RES-E compensated for by additional generation from CCS or nuclear power plants.

5.3.4. Strengthened Efficiency (EFF)

The electricity demand in the other scenarios decreases steadily until 2050. In the EFF scenario the peak in cumulated European electricity demand is reached in 2030. Afterwards, efficiency measures prevail and demand decreases until it reaches today's levels again in 2050. This scenario tries to explore how a significantly lower electricity demand would affect the optimal technological choices on the supply side. Naturally, the total costs of the EFF scenario will be significantly lower, but the same is also assumed for the average generation costs; a lower demand presumably allows smaller proportion of fluctuating RES-E in the supply mix, which might decrease the necessity for transmission and storage infrastructure.

As the model only covers the supply side endogenously, the costs for implementing energy efficiency and conservation measures cannot be calculated. Modelling the impacts and costs of efficiency measures on the numerous sources of electric-

---

[6]However, RES-E is not the only driver for which grid extensions are necessary; for example, it is also prerequisite for reaching the goals of the internal market for electricity, as discussed in section 2.2.

ity demand is a complex field beyond the scope of the applied model.[7] However, comparing the costs of this sector to those of others can lead to a first approximation of how much demand decreasing measures could cost for a costs-efficient decarbonisation.

The following sections will introduce all exogenous data used by the model in the scenarios. Differences between the scenarios will be highlighted.

5.4. Electricity demand and load profiles

Annual electricity demand is a major input parameter for the least-cost design of the power sector. As the comparison between the EFF and other scenarios will show in the next chapter, it influences many central results of the model. In this study, two different paths are applied: a high demand case, applied in all scenarios besides the EFF scenario, in which a low demand is assumed. The high demand case is the result of preliminary calculations of the *Netherlands Environmental Assessment Agency* (PBL) in the RESPONSES project. The data stems from a worldwide 450 ppm scenario calculated with the Integrated Assessment Model (IAM) *TIMER*. For an IAM, TIMER features a high level of detail in its modelling of energy demand and the impacts of efficiency polices. In the scenario, the developments in the power sector are characterised by the fuel switch towards electricity. The additional demand from the other sectors and the additional demand originating from GDP growth far outweigh efficiency and conservation measures.

The scenario's demand data is only available for the aggregated regions"Europe West" and "Europe Centre". Therefore, the aggregated demand was distributed using the proportions among countries of the 450 ppm scenario of the *ADAM* project[8], which follows a very similar storyline of ambitious decarbonisation. As it can be seen in table A.10 in Appendix A, in Central Europe, especially in the new MS, high GDP growth towards Western European levels in the course of increasing economic cohesion substantially drives energy demand. In turn, the growth in electricity demand is less strong in Western Europe, where efficiency measure are implemented earlier and more successfully and GDP growth is lower.

In the EFF scenario, a significantly lower demand is assumed, based on the *TRANS-CSP* study by the German Aerospace Center (see: DLR (2006)). The values had to be processed in some cases, because for some countries the data is

---

[7]On marginal abatement cost curves of energy demand see, for example, Fleiter et al. (2009) or Kesicki and Anandarajah (2011).

[8]For more information on the project, that is in many ways a predecessor of the RESPONSES project, please see: www.adamproject.eu.

available only in aggregated form (e.g. Benelux). In these case, the current proportion between the countries is kept throughout the scenario years. The resulting demand development is shown at country level in table A.11 in Appendix A.

Although the study was published before the economic crisis, the medium- and long term developments in energy demand developed within it seem plausible. In the scenario, electricity demand reaches its peak around 2030, after which demand decreases again until it reaches levels similar to today's. This seems realistic because previous experiences of decreasing demand suggest that enforcing rapid changes in energy consumption behaviour is difficult.

Alternative or additional scenarios in which demand developments are characterised by efficiency measures would also be worth exploring; this particular one was chosen as it seems ambitious but feasible. Analysing it with the developed model aims at generating a deeper understanding of how altered level of demand affect other variables of the system; the resulting trends are expected to be transferable to other scenarios with a similar success of efficiency measures.

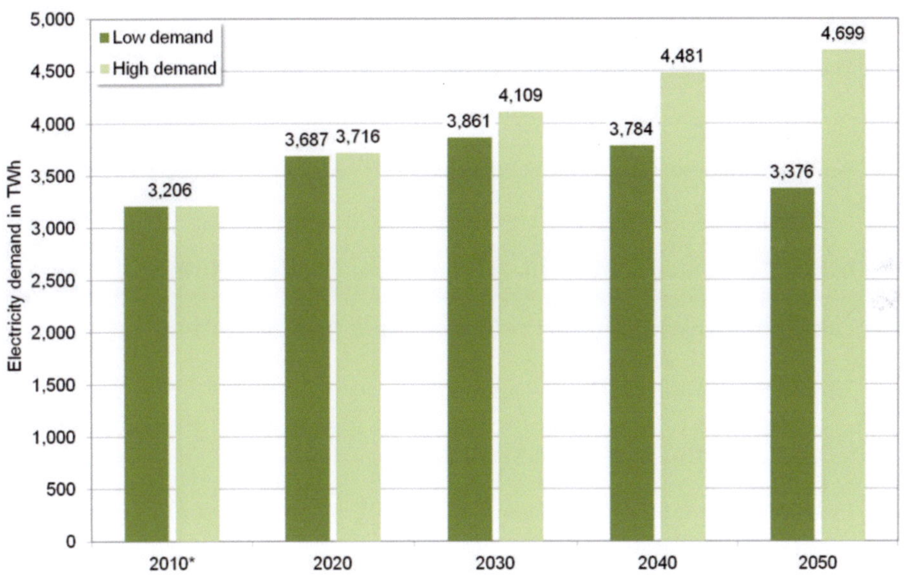

Figure 5.2.: Net electricity demand plus 6.5 % internal grid losses used as input in the scenarios. * 2010 values presented only as orientation, with data based on Eurostat (2012).

The developments in the two demand demand paths are depicted in figure 5.2. Values are given as net electricity demand plus *"internal grid losses"* of 6.5 %. In

PowerACE-Europe, internal grid losses are defined as transmission and distribution losses incurring either in the distribution grid or in the internal parts of the transmission grid not incorporated in the endogenous calculation. The applied value is based on calculations of Targosz (2008), who concludes that losses averages 7.3 % for the EU. Calculated endogenous grid losses are approximately 1 % for the early years of the scenarios, though they increase in subsequent years due to several reasons. Other losses, such as storage losses or self-consumption, are calculated endogenously during the model runs.

The hourly load profiles are generated in an iterative scaling procedure. The hourly load profiles of the year 2008 published on the Transparency Platform of the TSOs (ENTSO-E, 2012a) serve as a basis. The meteorological data, that defines the feed-in profiles for renewable energies also stems from 2008. That way correlations between weather and electricity demand are implicitly included. For example, a cold windy winters day will generate a relatively high electricity demand in countries with a large proportion of electrical heating systems, while a sunny summer day will be characterised by increased demand from air conditioning especially in Southern Europe.

For the following countries, data other than the ENTSOE-E profiles are used, because no data was available for 2008 when the input data was defined.

- United Kingdom (UK): data published by the TSO *National Grid* (National Grid, 2010)

- Ireland: data published by the TSO *EirGrid* (EirGrid, 2010)

- Estonia: data published by the TSO *Elering* (Elering, 2010)

- Malta: load profile of Italy is applied

- Cyprus: load profile of Greece is applied

- Lithuania: demand data is available from 2010 onwards, but for higher weather correlation the profile of Estonia is applied

- Latvia: demand data is available from 2009 onwards, but for higher weather correlation the profile of Estonia is applied

Unfortunately, the data published by ENTSO-E itself is subject to certain assumptions. Currently no market participant or authority has exact data on the total hourly electricity consumption of a country. This is largely because of industrial self-supply, i.e. companies producing at least a proportion of their own electricity, without feeding it into the grid. As it is not transported on the high voltage levels, it is not monitored by the TSOs. This results in certain inconsistencies in the available demand data: The sum of the hourly values for a particular month does not equal the monthly values and the sum of the monthly values does not necessarily equal the annual demand. This issue is discussed, for example, by Ellersdorfer et al. (2008). In PowerACE-Europe, it is dealt with by a method similar to the one presented by Ellersdorfer et al.: the hourly values are scaled to fit the proportions between the month but meet the annual demand.

In the scenarios of this thesis, the same demand profiles are applied in each year. This means that changes in the time of electricity consumption, for example due to new appliances, such as e-mobility, are not taken into account. Influences of altered load profiles appear to be a relevant field for future research.

5.5. Techno-economic assumptions on conventional power plants

The following sections will summarise the applied assumptions regarding the techno-economic parameters of conventional power generation technologies. The parameters have a critical impact on the modelling results; all of them influence the *levelised cost of electricity* (LCOE), which is often used for economic comparisons between technologies. However, in PowerACE-Europe, LCOE is calculated implicitly in the model and is only one component of the decision process.

Conventional power plants are well-known technologies that have been used for decades in large quantities. All the more surprising is the fact that in the literature a wide range of assumptions on their techno-economic parameters can be found. The main reason for this issue is pointed out by Finkenrath (2011, p. 16):

> "Several methodologies are used to estimate economic data [...]. There is neither a standardised methodology nor a set of commonly agreed on boundary conditions, which adds to the complexity of comparing data from different studies. Moreover, some factors are often not fully transparent, such as costing methodologies, sources of costs, the exact scope of data as well as assumptions on individual cost parameters."[9]

This represents a serious issue, as it places researchers in the difficult position of having to choose from a wide range of plausible parameters without biasing results. It can only be approached by comparing existing studies on the topic albeit knowing that the authors were faced with the same issue.

---

[9]Although the paper focuses on techno-economic parameters of CCS power plants, the critique applies to the discussion of other technologies' parameters as well.

To the knowledge of the author, the most recent and comprehensive compilation of scientific analyses dedicated to the topic of LCOE and the underlying techno-economic assumptions of conventional and renewable power plants is gathered by the Energy and Ecology Blog (2012); it presents a collection of 45 publications from 2008 to 2012. As most techno-economic parameters influence LCOE, the analyses cover all aspects relevant to the parameters in this work. Besides work specifically dedicated to power plant parameters, most long-term scenario studies also publish the underlying assumptions.

Since on many parameters no consensus exists, updating the techno-economic parameters of the model is a continuous task of comparing sources. In doing so, many of the publications discussed in the Energy and Ecology Blog (2012) were evaluated and compared. For conventional power plants the following sources have to be highlighted either for their influence on the parameters chosen for this thesis or because they are not included in the collection studies outlined above.

- The work "Projected Costs of Generating Electricity" (IEA, 2010) is not only very rich in details on power plant parameters, it also provides a regional dimension. This is because power plant prices differ substantially between the world regions and even countries. The work presents a detailed sensitivity analysis on the impact of key parameters on the LCOE.

- The Appendix to the Study "Roadmap 2050" [10] (ECF, 2010) includes detailed assumptions on power plant parameters.Similar parameters are applied in this thesis especially for gas power plants, both open- and combined-cycle.

- The series of Greenpeace scenario studies "energy [r]evolution" (the latest publication being Teske, Muth, et al. (2012)) is not included in the LCOE discussion, but presents plausible developments for both conventional and RE power plants.

- Recently, Blesl et al. (2011) presented LCOE calculated using Monte-Carlo simulations. Instead of using a fixed parameter set, the authors apply stochastic distributions to central parameters, such as specific investments and efficiency. This seems to be a valuable approach that moves away from applying one fixed development for every techno-economic parameter.[11]

---

[10]This part, Appendix A, is no attached to the main report, but can be downloaded at: www. roadmap2050.eu/downloads.

[11]However, the approach itself is not applicable in the model PowerACE-Europe.

Special attention has to be paid to the parameter selection regarding CCS power plants. As of 2012, no CCS power plant of commercial size is in operation; consequently, the uncertainties regarding the techno-economic parameters are even larger than for other technologies. The issue is complicated further by the fact that three different CCS technologies exist: Pre-combustion, post-combustion and Oxy-fuel combustion (the differences between the technologies are discussed, for example, in: Wuppertal Institut (2010, pp. 69-86). The wide range of costs assumptions, even for specific CCS technologies, is evidenced in the meta-study of Finkenrath (2011). For example, for the Oxy-fuel carbon capture from coal-fired power generation, specific investments from 2,875 EUR/kW to 5,106 EUR/kW have been calculated by different studies. Both values are stated for plants in OECD countries, while for China the costs are significantly lower.

Due to the lack of information on the real future costs, data on specific investments and O&M in the work at hand are based on the average premium compared to the respective non-CCS power plants calculated by Finkenrath. Lower than average costs are chosen for CCS CCGT power plants, as a study by the Wuppertal Institut (2010, p. 220) indicates that significantly lower cost are possible. The influence of this choice is tested with a sensitivity analysis in section 6.3.

Similar uncertainties can be found for nuclear power plants; although much more experience exists in nuclear technology, most of the power plants were built in the 1970s and 1980s. Previous cost data has become outdated, for example because security regulations have become much stricter. The lack of recent representative data makes cost estimates for future nuclear reactors difficult. The recent experiences with the *European Pressurized Reactor* (EPR) technology, a third generation nuclear power technology, have been rather disappointing: The estimated costs of the EPR in Finland more than doubled from the originally planned 3.2 billion EUR to the current approximately 7 billion EUR, according to prognoses gathered by Greenpeace (Greenpeace, 2012). The additional costs are to some extent based on the construction being substantially behind schedule. Similar issues exist for the EPR in Flamanville in France, for which the specific investment increased from the originally planned 2,024 EUR/kW to approximately 3,700 EUR/kW. The plant is still at least three years from completion (Nuclear Engineering International, 2011). However, the high costs could be based on the fact that the EPR technology is new and no nuclear power plants have been built in Europe for over 20 years.

A meta-study by Severance (2009) concludes LCOE of 25 - 30 US-cents/kWh for nuclear energy. This seems rather high, as even lower LCOE of 150 EUR/MWh are possible at investments of 7,000 EUR/kW[12]. In this thesis, a "medium level" of specific investments is applied. Costs are not changed over the scenario years, representing a compromise between the cost increases over recent decades and the future learning potential. However, the broad range of possible costs is accommodated in a sensitivity analysis (see section 6.3).

The chosen power plant parameters are depicted in table 5.2. As mentioned, every figure in this table is arguable and differing values are given by other studies. However, to the best knowledge of the author, no chosen parameter is extreme, neither too optimistic nor pessimistic. The chosen parameter set implicitly reflects the "optimistic" background of the decarbonisation scenarios. However, because the model does not include lowered efficiencies in partial load the chosen efficiencies are lower than the maximum efficiency possible with the respective technology as an approximation.

Furthermore, the actual input table in the model contains a number of other power plant technologies, such as integrated gasification combined cycle power plants and other advanced, non-CCS technologies using fossil fuels. Even applying very optimistic assumptions on cost parameters, none of the technologies becomes relevant in the model results in any decarbonisation scenario; consequently, the technologies are omitted in the following analysis.

5.5.1. Assumptions on fossil fuel and EUA prices

The LCOE of the conventional technologies are determined by the power plant parameters, the fuel price developments and, in the case of fossil fuels, carbon emission prices such as EUA prices. The price paths assumed in the following in all scenarios are depicted in figure 5.3. The price paths for coal, lignite, oil and gas as well as for EUAs stems from preliminary calculations of PBL in the 450 ppm scenario of the RESPONSES project (see: Deetman et al. (2012)), i.e. the same source as the electricity demand in the "High" demand scenarios. Prices of lignite[13] are kept at today's level. For nuclear, a price increase of 50 % by 2050 is assumed.

The prices reflect the worldwide developments in the scenario of PBL: The demand for coal increases by 35 % between 2020 and 2050, since CCS coal power plants are a major mitigation technology in the worldwide scenario thus driving up demand. Prices for natural gas increase by 70 % for the same reason. Oil prices

---

[12] Assumed interest rate: 10 %, fixed annual O&M of 80 EUR/kW, utilisation of 7,000 h/a.

[13] The term "price" is not really fitting for lignite, as the fuel is hardly traded or transported due to its low energy density. The price of lignite should be seen as estimated costs for mining lignite and transporting it to the power plant.

110

Table 5.2.: Assumed developments of the techno-economic parameters of conventional power plants for the scenario years.

Technology	Year	Electric efficiency	Life-time	Specific investment	$O\&M_{fix}$	Carbon capture rate
Unit	-	[%]	[a]	$[\frac{EUR}{kW}]$	$[\frac{EUR}{a \cdot kW}]$	[%]
Nuclear	2020	35.0	40	3,500	80	0
Nuclear	2030	35.0	40	3,500	80	0
Nuclear	2040	35.0	40	3,500	80	0
Nuclear	2050	35.0	40	3,500	80	0
Coal	2020	46.0	40	1,300	30	0
Coal	2030	46.3	40	1,283	30	0
Coal	2040	46.7	40	1,267	30	0
Coal	2050	47.0	40	1,250	30	0
Lignite	2020	44.0	40	1,600	40	0
Lignite	2030	44.3	40	1,550	40	0
Lignite	2040	44.7	40	1,500	40	0
Lignite	2050	45.0	40	1,450	40	0
Gas (CCGT)	2020	58.0	30	864	15	0
Gas (CCGT)	2030	59.0	30	826	15	0
Gas (CCGT)	2040	60.0	30	788	15	0
Gas (CCGT)	2050	61.0	30	750	15	0
Gas (GT)	2020	40.0	30	400	15	0
Gas (GT)	2030	40.7	30	383	15	0
Gas (GT)	2040	41.3	30	367	15	0
Gas (GT)	2050	42.0	30	350	15	0
CCS CCGT	2020	50.0	30	1,149	20	85
CCS CCGT	2030	53.0	30	1,074	20	88
CCS CCGT	2040	55.0	30	1,001	20	91
CCS CCGT	2050	56.0	30	938	20	95
CCS Coal	2020	37.0	40	2,275	40	85
CCS Coal	2030	37.7	40	2,246	40	88
CCS Coal	2040	38.3	40	2,217	40	91
CCS Coal	2050	39.0	40	2,188	40	94
CCS Lignite	2020	35.0	40	2,800	40	85
CCS Lignite	2030	35.7	40	2,713	40	88
CCS Lignite	2040	36.3	40	2,625	40	91
CCS Lignite	2050	37.0	40	2,538	40	94

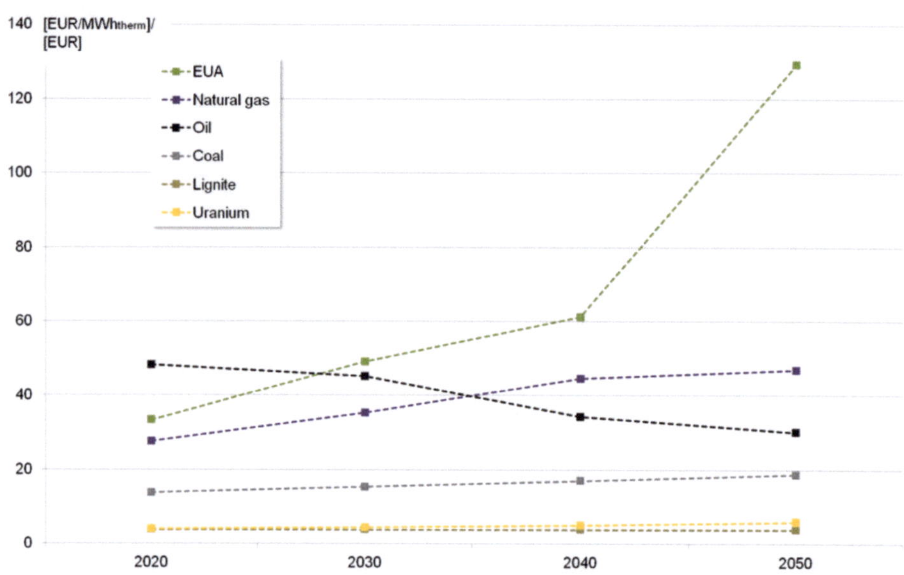

Figure 5.3.: Prices for fuels (in EUR/MWh$_{\text{therm}}$) and for EUAs (in EUR) applied in all scenarios. Source: Deetman et al. (2012) and own calculations.

decrease, as strong mitigation policies decrease demand from the transport sector. Whether or not such a price decrease is feasible can be questioned; however, the oil price has very little impact on the developments in the power sector if prices are exogenous; the construction of new oil power plants is not possible in the model and the low efficiency of most existing units means they become very rarely used peak load power plants.

The developments in fossil fuels prices are to some extent overlaid by the strong increase in EUA prices, especially in the later years. In consequence the share of the fuel expenditures on the total LCOE decreases, whereas EUA expenditures become a major cost component. The calculations of the worldwide developments are based a normative scenario approach, analysing the question of the conditions necessary for a meeting the 2° C target. The high carbon price is the result of the immense level of effort needed to keep this target achievable. This becomes very visible towards the end of the scenario horizon, with CO_2 prices reaching almost 130 EUR/t in 2050.

The power sector is responsible for a large share of European fuel consumption and even represents a significant part of world fuel demand. Therefore, in reality, the demand for fuel, which is calculated by PowerACE-Europe endogenously, would impact fuel prices. The interdependencies are probably stronger for gas, for which the markets are more regional than for coal which is traded over large distances. Consequently, treating fuel and carbon prices as exogenous is a simplification. However, although differing fuel and carbon prices could presumably increase consistency within the scenarios, they would decrease the interpretability of the results; with several core parameters being altered simultaneously, existing cause-and-effect interrelations would not be comprehensible. Furthermore, with the increase in RES-E in the power sector and the decreasing utilisation of power plants, the impact of fuel prices on behaviour decreases over time.

5.6. Renewable energy power plant parameters

The economic parameters of RE technologies have to be defined in the model for the endogenously calculated options, i.e. wind and solar power technologies. The uncertainties regarding the parameters of renewable energy technologies are in some aspects larger than for conventional power generation. Because renewable energies are currently subsidised, the observable market prices are to some extent biased by the respective support policies (Hearps and McConnell, 2011). Because most RE technologies are at an earlier stage of development, technological learning causes costs to decrease faster then for mature conventional technologies. Furthermore, learning often takes place by leaps and bounds in the early stage of the technological life-cycle. It has to be taken into account that, on the one hand, previous attempts to forecast the developments of RE technologies tended to, often significantly, underestimate the speed of learning. On the other hand, the costs of some technologies, especially wind power, are influenced to a high degree by commodity prices, e.g. steel and copper (Panzer, 2012). A rise in commodity prices can outweigh technological learning and lead to net increase in specific costs per MW.

Consequently all approaches to forecast future parameters of RE technologies come with substantial error margins and should be interpreted accordingly. Furthermore, it means that estimations are often quickly outdated, for example, because predicted future costs are already undercut by today's costs.

Nonetheless, several recent studies produce comparable ranges of parameters for wind and solar technologies for the medium- to long-term future. A notable example of these studies is the extensive study on RE technologies by the IPCC (see: IPCC (2012)) or the series "Renewable Energy Cost Analysis" by the International Renewable Energy Agency (IRENA) (see: IRENA (2012c,b,a). Recent data can also be found in the "Transparent Cost Database"[14] of the National Renewable Energy Laboratory (NREL).

For application in the model, the economic assumptions provided by Zickfeld and Wieland (2012) are chosen. This study includes a consistent set of parameters that is up-to-date and in the ranges discussed in the studies above for most technologies. One of the tasks in this thesis is to analyse the optimal share of RES-E for decarbonisation strategies under different circumstances. Extreme cost assumptions, either very optimistic or pessimistic, are likely lead to foreseeable results in the optimisation model. Still, the chosen economic parameters are provided in table 5.3. The specific investments of CSP given in the study seem to be relatively optimistic, though possible.

Table 5.3.: Economic assumptions regarding the four endogenously calculated RE technologies, based on Zickfeld and Wieland (2012).

Technology	Year	Life-time	Specific investment	$O\&M_{fix}$
Unit	-	[a]	$\left[\frac{EUR}{kW}\right]$	$\left[\frac{EUR}{a \cdot kW}\right]$
Photovoltaics	2020	25	1,000	35
Photovoltaics	2030	25	830	30
Photovoltaics	2040	25	760	25
Photovoltaics	2050	25	700	19
CSP	2020	30	3,300	74
CSP	2030	30	2,500	64
CSP	2040	30	2,250	55
CSP	2050	30	2,000	45
Wind (onshore)	2020	25	1,100	20
Wind (onshore)	2030	25	1,000	20
Wind (onshore)	2040	25	930	20
Wind (onshore)	2050	25	900	20
Wind (offshore)	2020	20	2,000	80
Wind (offshore)	2030	20	1,650	66
Wind (offshore)	2040	20	1,500	60
Wind (offshore)	2050	20	1,340	54

The capital costs are annualised and combined with the data of the regional potentials. The available potential is calculated by a GIS model, taking various aspects into account, for example available area, nature conservation, wind speeds, hub heights and wind power curves of exemplary turbines. For the cost-potential steps in this thesis, a relatively low spatial resolution is chosen, as a high number of

---

[14]See: http://en.openei.org/wiki/Transparent_Cost_Database.

potential steps increases calculation time to unacceptably high levels. The resulting cost-potential curves are depicted in figure 5.4. As it can be seen especially in the case of photovoltaics, the applied criteria regarding land use are restrictive thus limiting the potential. For example, only one percent of agricultural areas can be used for solar modules. The strictly technical potential of all the technologies is significantly higher.

The relative decrease in LCOE is strongest for photovoltaics, although the learning rates are significantly below the ones observed over the most recent decade. The slower learning is based on the fact that previous cost decreases were concentrated on the costs and prices of the photovoltaic modules, which accounted for the largest share of total costs. After the decrease in module prices, other cost components, such as inverters, installation frames and labour costs, are growing in relevance; for these, cost cannot be decreased as fast as for the modules.

Similar strong cost reductions are foreseen also for offshore wind power; the technology is currently at an early stage of diffusion, but with ambitious plans by 2020. High reductions in LCOE seem possible especially in the first 20 years.

In contrast to the learning potential of offshore wind power, costs decrease only slowly for onshore wind power. The technology is already very mature and the potential for cost reductions are to some degree already exploited. For example, increases in turbine size and rotor diameter, which played a important role in previous costs reductions, cannot be extrapolated into the future. With current materials and designs, upscaling seems to be close to the point at LCOE cannot be decreased much further (see for example: Fink (2011)).

The costs reductions for CSP are high, because the scenarios in Zickfeld and Wieland (2012) describe a world with high deployment of CSP technologies, especially in Northern Africa and the Middle East, which drives down costs. Furthermore, the costs of CSP depicted in figure 5.4 do not include storage losses, i.e. assume that solar energy is converted to power without storage processes.[15]

With the chosen parameters, onshore wind power continues to be the RE technology with the lowest LCOE. This means that offshore wind power does not become competitive with onshore sites from a purely cost based perspective and will consequently be built by the model only for portfolio reasons.[16] To some extent, this is

---

[15]The amount of power that is stored is a model result and cannot be calculated before the actual runs. As these losses drive up the cost, the implicit "net LCOE" are higher in the model.

[16]At a certain point, increasing the capacity of a fluctuating technology in a country is no longer efficient due to the profile assumed to be homogeneous for each country and RE technology. Therefore, balancing the onshore wind capacities with other technologies, such as offshore wind, can be cost-efficient, even if the LCOE of the technologies are higher.

the result of the low spatial resolution of the potentials; if wind speeds are averaged over large areas, the most attractive sites "disappear".

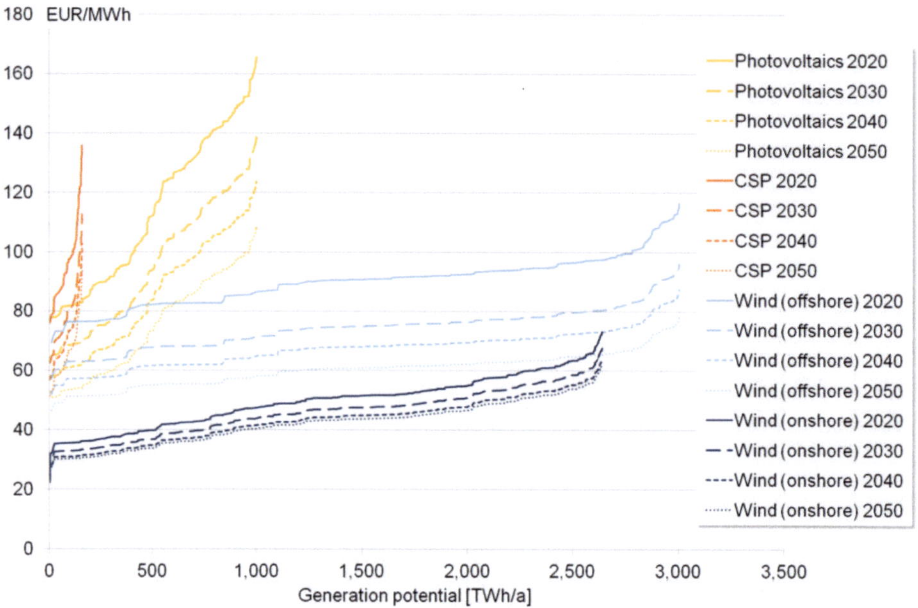

Figure 5.4.: Cost potential curves for wind and solar power applied in the scenarios.

The meteorological profiles by 2008 are applied for all RE technologies. The profiles are scaled to match the average generation at the sites. The year is chosen for the high data availability for both RE profiles and electricity demand.[17]

5.7. Renewable energy diffusion scenarios

The development of renewable electricity generation capacity until 2020 is based on the NREAPs as published by the European Commission (2011e). Although in reality the developments will deviate to some extent from the target paths, it is assumed that the paths are more or less met, as governments will take measure to ensure target fulfilment. After 2020,the diffusion path of the exogenous RE technologies, i.e. all technologies besides wind and solar power, are defined with the model *PowerACE-ResInvest*. This agent-based model, which is described in detail

---

[17]The chosen weather year has an moderate impact on the overall results if the energy output remains the same, i.e. if the same generation is scaled with a different profile. A lower generation however, e.g. from a "bad" wind year has an impact. The deterministic approach of the model currently does not allow stochastic optimisation.

in the doctoral thesis of Held (2011), simulates the diffusion of renewable energies up until 2050. The model contains detailed techno-economic data on specific investments, learning rates and generation potential for RE technologies in Europe. It differentiates 14 generation technologies in more than 5,000 potential steps. The diffusion process is modelled from the perspective of investor agents, which pursue their respective rationale by evaluating potential sites and national support schemes. The latter are included for the current support scheme, for example, as FIT or quota with TGC trading, and are adjusted by "policy agents" if national targets are over- or underfulfilled. The model also includes technological learning and simulates the expansion of construction capacities for the different technologies.

To match the ambitious goals in the scenarios, the policy agents are modelled with ambitious goals regarding RE. This means that the countries aim at very high proportions of RES-E and adjust their policies accordingly. It is, of course, impossible to forecast the measures that policy makers will implement over the coming four decades. Hence, results should be interpreted as a scenario rather than forecasts. Furthermore, it has to be pointed out that the results of the simulated investments in renewable energy technologies differs significantly from a pure least-cost approach; the simulation results in a rather distributed allocation of all types of RES-E plants across Europe.

Although the only relevant results from PowerACE-Europe are those regarding the technologies that are not calculated endogenously, the ResInvest model simulates all technologies simultaneously. Wind and solar power are included in the simulation, as they influence the national target fulfilment monitoring and policy adjustment processes. The resulting power generation in the scenario years for all technologies is given in table A.1 in Appendix A.

As it can be seen in figure 5.5, the growth of exogenous technologies between 2020 and 2050 is moderate. Hydropower grows insignificantly since the potential for large-scale plants is almost completely depleted and small-scale sites are not very attractive for investors under the current support schemes.[18] Similar issues hinder the expansion of tidal power, for which the geographical conditions, e.g., a bay with a narrow mouth and a high tidal range, are limited to a few sites in Europe. Electricity generation from biowaste, landfill and sewage gas also has small potential for growth beyond 2020.

---

[18] As the prospects for future cost decrease seem rather low for the technology, policy makers do not set high incentives for it. However, in future electricity systems small-scale hydropower might gain interest as a decentralised, dispatchable technology.

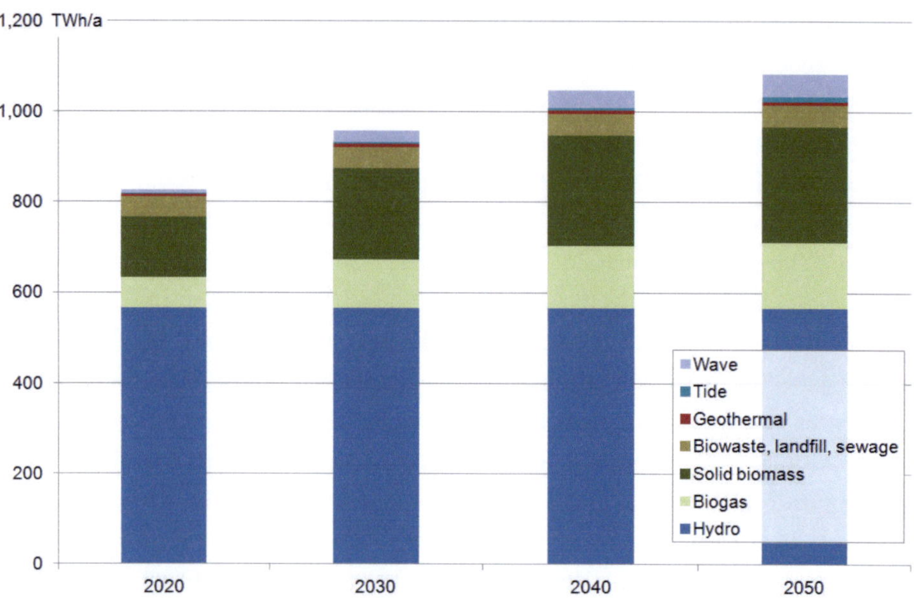

Figure 5.5.: Annual generation from the RE technologies that are exogenous to PowerACE-Europe, calculated with the RES-E diffusion model ResInvest. The data covers the EU-27+2.

In the case of geothermal energy it has to be noted that ResInvest only covers conventional geothermal power, i.e. utilising high temperature water flows near the surface, which is possible in only a few regions in Europe. Enhanced geothermal systems (EGS), e.g. the hot dry rock technologies, which are theoretically possible in many more regions, are not included. However, ongoing discussions about technical issues, especially induced seismicity (cf. Bundesverband Geothermie (2010)), and high costs render a rapid diffusion of EGS questionable.

For the other technologies, lack of additional potential is not the central issue. Electricity generation from wave power has a huge generation potential: Krewitt et al. (2009) estimate a worldwide technical potential of over 5,000 TWh per year. However, as the authors also point out, costs are currently among the highest of all RE technologies. In ResInvest the costs of specific investments decrease through learning curves, and so the model cannot predict technological leaps. This means that although costs of tidal power decrease, they do not come down fast enough for a real breakthrough of the technology. The observable diffusion is the result of the currently generous support schemes in the UK and Portugal, which the model extrapolates into the future.

The strongest growth anticipated by the model for is biomass and biogas; the generation from these almost doubles between 2020 and 2050. Generation from solid and gaseous biomass increases to 256 TWh and 145 TWh, respectively. However, it is clear that such an increase in biomass utilisation for energy purposes is subject to questions of sustainability. Expanding the energetic use of biomass concerns biodiversity issues and indirect land-use change, as well as moral questions, for example, concerning due to the interactions with the food sector. The potentials for both solid and gaseous biomass in the ResInvest model reflect these concerns and are limited to levels that are assessed to be in line with sustainability criteria. [19] However, up until 2050, the available biomass potential is almost completely utilised in the scenario.

5.8. Other scenario assumptions and limitations

A further exogenous restriction of the solution space of the scenarios is the capping of the annual CO_2 emissions. The applied values are given in table 5.4. Introducing the cap serves two purposes. Firstly, it ensures compliance with the definition of decarbonisation scenarios in section 5.1; a gradual approach would otherwise have to be applied, iteratively increasing carbon prices until the desired emission reduction was achieved.

Table 5.4.: Annual CO_2 emission caps applied in scenarios.

Scenario year	CO_2 cap
2020	700 Mt
2030	400 Mt
2040	150 Mt
2050	75 Mt

Secondly, capping emissions in every year, with gradually decreasing values, results in a "smoother" transformation; if only the emissions are only restricted in 2050, the model could theoretically produce a carbon intensive power plant park and generation mix for early years. Then, over the last 10 years, the model could change rapidly to clean technologies to meet the carbon cap. This approach would not be feasible in reality or at least would be very costly. The model underestimates the costs of such an approach. Among other costs not included in the model, the exogenous techno-economic parameters of the clean technologies, for example CCS

---

[19] See: Held (2011, 82pp); the author relies on a number of studies, e.g. the sustainability assessment of Wiesenthal et al. (2006), to derive the available potential of biomass and biogas.

and the RE technologies, implicitly assume ambitious diffusion and would otherwise be more expensive in 2050.

5.9. Critical reflection and conclusion

This chapter introduced the rationale of the scenarios that will be analysed in the following chapter and described their applied input parameters. The definition of the long-term scenarios represents a trade-off between defining consistent and coherent developments, e.g. taking into account the interdependencies between fuel consumption and prices, whilst keeping the differences between the scenarios at levels at which changes in results can be traced to causes. For the scenarios in this thesis, the latter is given greater importance because this work seeks a better understanding of the relationship between the technological components of the electricity system. The results of the model runs should be understood accordingly and not be mistaken for a forecast.

Naturally, for all input parameters, values other than the ones applied here can be found in the literature. The chosen assumptions represent optimistic, though not extreme, technological and political developments in the future. The impacts of changes in selected key parameters are covered through sensitivity analyses.

A certain shortcoming of the input data is the uniformity of the load profile over the scenario years. In reality, the shape of the load profile will clearly change, but the detailed influences are unknown. Measuring the impact of altered load profiles should be examined in future research. Similarly, it seems worthwhile to apply different meteorological years for the scenarios years. Although this is already possible with the current model, it still lacks the tools for a systematic analysis of the impacts. For example, analysing the shadow prices of RE feed-in profiles could reveal which weather conditions are most challenging and how these affect the overall solution.

6. Results and conclusions

In the previous chapter, four scenarios were defined. The results of the model runs, calculated with the model PowerACE-Europe, will be discussed in this chapter. It has to be borne in mind that each run creates a large amount of data consisting of several hundred million variables; distilling the core results from these and presenting them in textual form is challenging. Usually, observations and interpretations or conclusions should be separated as far as possible; in this particular case, a strict separation is difficult. For example, consider a model result where the installed capacity of nuclear power decreases significantly by 2050, which taken for itself is just an observation. However, the fact that under the scenario's circumstances nuclear power faces a difficult market environment is also a conclusion. Furthermore, it has to be considered developments have to be analysed over time for a high number of variables (such as installed capacities, utilisation, curtailment and costs) and many different technologies in four different scenarios. Therefore, a strict separation of observation and interpretation into different sections does not seem practicable. In this work a different approach is chosen, in which the conclusion from the presented findings concerning a certain aspect is given in the same section. Interpretations of the results that are ambiguous are marked as such.

In the discussion of the results, the "Optimistic decarbonisation" (OPT) scenario will act loosely as a point of reference. Firstly, the results of this scenario will be introduced and discussed in detail in section 6.1. After this, substantial deviations between the other scenarios' results and the OPT case are discussed in section 6.2. By discussing one scenario in detail, the functioning of the model and the complex interdependencies between the model system components can be observed. This is done simply to facilitate comprehension and should not be interpreted as a valuation between the scenarios.

To avoid repetitions and ambiguities, a development observed without explicitly stating dates, e.g. "The RES-E share grows from X to Y" refers to the period between 2020 and 2050. Nevertheless, it has to be borne in mind that the scenario horizon covers the years 2011 to 2050.

6.1. Developments in the OPT scenario

The OPT scenario is well suited to be discussed first, as it has the highest degree of freedom for the model; the least-cost solution can be determined without the limitations of the "No CCS" (NoCCS) or "Hampered Grid" (GRID) scenario. Therefore, the principle mechanisms of the model can be observed. The "Strengthened Efficiency" (EFF) scenario deviates in its demand and is thus, by definition, very different from the other scenarios.

6.1.1. Power generation mix and installed capacities

In order to understand the results of the model runs, the installed capacities and their utilisation are of particular interest because they are affected by virtually all model parameters. Figure 6.1 depicts the generation mix in the OPT scenario. As additional information, losses of interconnections and storage facilities as well as curtailment are shown below the horizontal axis. Subtracting these from the total power generation equals the electricity demand including interior losses of the scenario. In turn, figure 6.2 shows the generation installed capacities during the scenario years. The installed capacities of storage facilities are included in the figure, although storages are not able to generate electricity from primary sources. However, in combination with fluctuating generation, they can behave in a similar way to dispatchable power plants and provide backup capacity. The figures for both diagrams are also given in table A.2 and A.3 in Appendix A.

The OPT scenario is characterised by an extensive restructuring of the power sector, more precisely of the supply side of the power sector and the necessary infrastructure. The most obvious and fundamental change is the strong growth of RES-E: The *total RES-E generation potential*[1] increases from 1,713 TWh in 2020 to 3,869 TWh in 2050. The *net RES-E share in generation*, defined as

$$\text{Net RES-E share} = \frac{\text{RES-E generation potential} - \text{total curtailment}}{\text{Total electricity generated}} \tag{6.1}$$

increases from 45.3 % in 2020 to 79.0 % in 2050. Consequently, the share of conventional power decreases to 21 % which is almost entirely generated by CCS power plants. Below the horizontal axis, the increase in losses and curtailment can be

---

[1]In the following, "RES-E generation potential" will refer to the potential generation from RES-E plants not taking into account the generation that has be curtailed in situations of oversupply. The model does not define which RE technology would be curtailed, as in reality this depends on several factors.

122

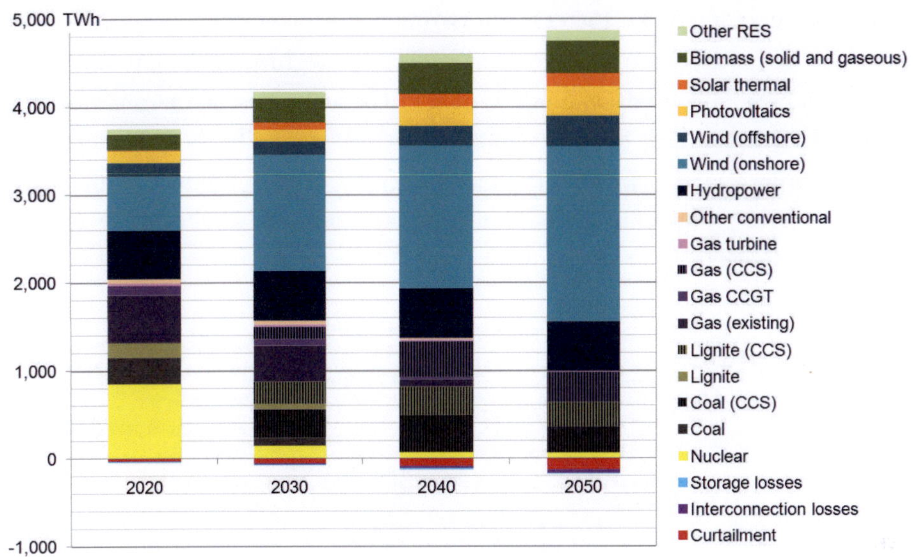

Figure 6.1.: Electricity generation, losses and curtailment in the OPT scenario.

observed. In 2050, 3.4 % of RES-E generation potential has to be curtailed because its utilisation would be too costly (see section 6.1.6).

The relationship between the RE technologies shifts: in 2020, one third of RES-E generation comes from hydropower, one third from onshore wind and one third from other RE technologies. In 2020, power generation from onshore wind in many countries exceeds the targets of NREAPs; in total, the onshore wind targets of EU MS are greater than 108 TWh: Norway and Switzerland generate additional 4 TWh. However, the growth of the other RE technologies is dominated[2] by the NREAPs, i.e. targets of the NREAPs are met but not surpassed.

The growth in conventional power generation starts very moderately: in 2020, only 131 TWh are generated in newly built gas power plants, both open-cycle and CCGT. This demonstrates the low degree of freedom within the model up to 2020 as the NREAPs and the existing power plants largely determine the developments until 2020. Interestingly, the model does not yet invest significantly in CCS power plants, although the technology is (optimistically) assumed to be available at relatively low cost.

The developments between 2021 and 2030 are characterised by a high necessity

---

[2]"Dominated" in this context means that the exogenously enforced value for a variable is at least as high as in a , ceteris paribus, freely determined model calculation.

to invest; in this time-frame 155 GW of existing power plants reach the end of their lifetime, while at the same time demand continues to grow. This gap is filled by both renewable and conventional power plants. Capacity installed in onshore wind continues to grow at high average annual growth rates of 8.8 %/a. Furthermore, the model utilises CSP plants for the technology's ability to produce dispatchable yet emission-free power. The generation from both offshore wind and PV remains at the NREAP level.

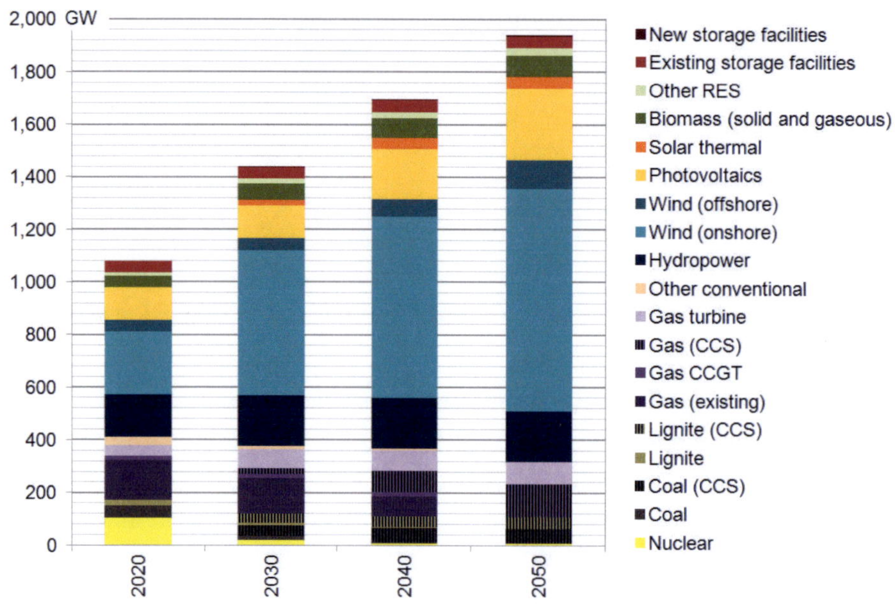

Figure 6.2.: Installed generation capacities in the OPT scenario.

Furthermore, the model starts to invest in all three available CCS power plants types. The different plant types take a similar place in the power system and the merit order as their non-CCS counterparts do today. Lignite power plants are limited to existing sites and capacities. The model replaces retiring lignite power plants with their CCS counterparts, which are also operated as base-load power plants; they compensate for their high capital costs through low variable costs. Coal-fired CCS plants are similarly used, i.e. with a high utilisation, whereas CCGT plants start at high utilisations, but move to the midload segment over time (see figure 6.4). Furthermore, nuclear capacity is increased by a total of 7.4 GW in the UK and Finland.

Between 2031 and 2040, the trends of the previous decade continue, especially for conventional power generation. However, the capacity of all endogenous RE technologies is increased, including offshore wind and PV. While the latter increases by 2040 only in the southernmost countries, offshore wind turbines are installed in most countries that have a coast, with the largest increase taking place in the North Sea.

By 2050, the electricity mix has changed immensely from that of today with 54 % of the generated electricity coming from fluctuating RES. Wind power generates the highest proportion of this, with onshore and offshore sites having potentials of 1,985 and 351 TWh, respectively. Flexible RES-E is provided by solar thermal power plants as well as the dispatchable part of biomass and reservoir hydropower; these technologies provide 906 TWh in total, which equates to 18.5 % of total electricity generation. PV generation potential increases to 333 TWh, which is 2.5 times the generation planned in the NREAPs.

However, the proportion of PV in power generation is surprisingly moderate; the reason might lie in the periodicity of solar energy. Figure 6.3 shows the sum of the generation potential in all hours and countries depending on the hour of the day. As it can be seen, the wind sites utilised by the model tend to produce slightly below average between morning and noon. This is complemented partially by the installed photovoltaic modules. Please keep in mind that figure 6.3 includes winter months as well as the countries in Northern Europe. Therefore, the proportion of PV on total generation on a sunny day in southern Europe is significantly higher and often requires curtailment. Increasing PV capacities beyond this point would disproportionately increase curtailment. A further diffusion of PV is also limited by the role that wind power has to play: in a power system based largely on wind power, which has a stochastic behaviour but weak periodicity at most sites, the role of technologies with a periodic profile depends on how well its periodicity fits the shape of the demand curve. Another way to put it is that PV only makes sense because demand is higher throughout daytime. As power generation from wind has a similar expected value in each hour, PV is only economic to the extent that demand is higher during the sunshine hours.

Only 21 % of total power generation in 2050 comes from conventional power plants; of this, 93 % is generated in CCS power plants. The proportion of power generation from open-cycle gas turbines not equipped with carbon sequestration is only 0.11 %, although it provides 4.4 % of the installed capacity. In order to under-stand the technological choices and their utilisation in the model it is worthwhile considering the utilisation or capacity factors.

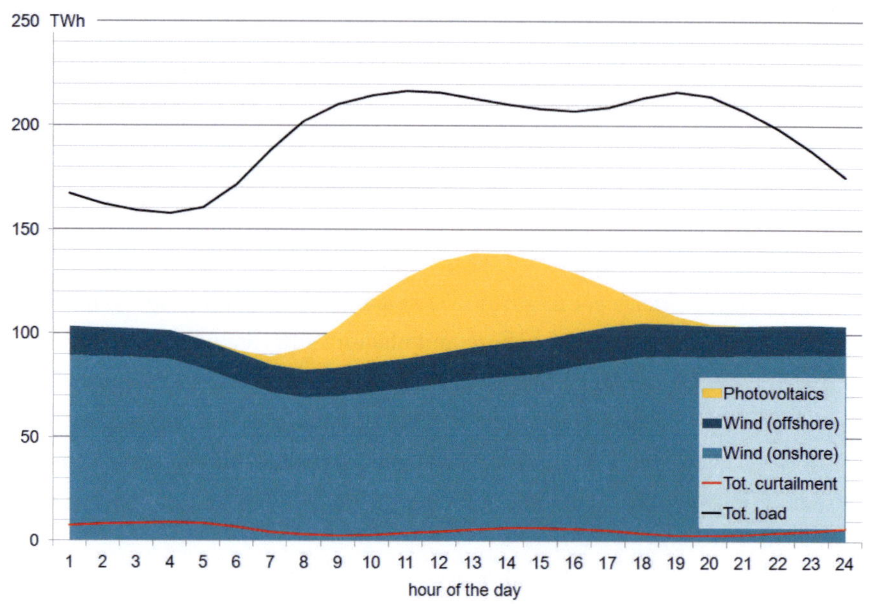

Figure 6.3.: Total electricity generation potential of fluctuating RE technologies, curtailed energy and load in 2050 of the OPT scenario in the hours of the day. The values are the sum of the generation from the respective technology throughout the whole year.

6.1.2. Utilisation and capacity factors of generation capacities

Figure 6.4 depicts the capacity factors of the endogenous conventional and RE technologies over time. As it can be seen, nuclear power plants demonstrate the highest utilisation, being close to the technically possible maximum.[3] This is basically the only situation in which nuclear power plants are built by the model: the capacity factor has to remain at a very high level throughout the whole lifetime of the plant. In many regions, the increase in fluctuating RES-E pushes base load power plants generation out of the market, because it decreases the number of hours in which electricity has a significant value. This development has the highest impact on nuclear power, as it is solely a baseload technology. Nevertheless, the effect can only be observed for all conventional technologies. The utilisation of both lignite and coal power plants is also high, but decreases in the later years[4]. Nevertheless, at 6,500 FLH in 2050, lignite remain the baseload power plants.

---

[3]The utilisation in 2020 is shown dashed, as no new nuclear power plant is built up to this point.
[4]However, installed CCS capacities in 2020 are small (below 1GW) and are consequently not very expressive.

126

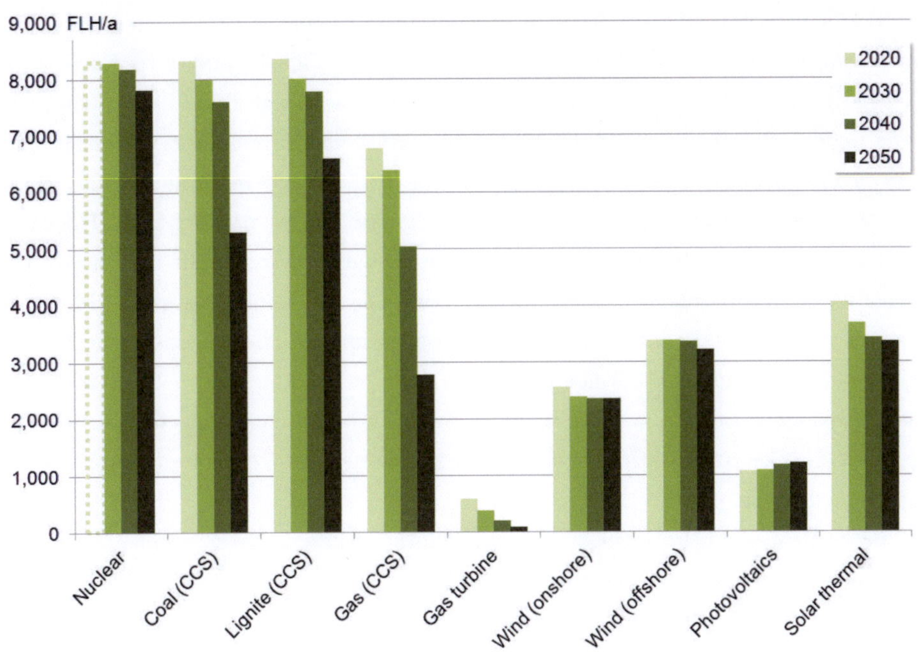

Figure 6.4.: Utilisation, expressed in full load hours (FLH), of the endogenous technologies over time in the OPT scenario.

At this point it should be noted that with the advancing diffusion of RES-E, load segments are increasingly linked to the residual load, after the generation from undispatchable RES has been subtracted, rather than the total system load. In academia, there is a broad consensus on this and the fact that a significant diffusion of fluctuating RES-E decreases the size of the baseload segment. However, the model results reveal that this development is significantly counterbalanced by the expansion of the interconnection capacities; although a particular country might have a negative residual load, demand might exist in other countries in the region. When analysing the possible utilisation of dispatchable power plants, the *regional residual load* gains importance over time. Its definition is fuzzy however,and cannot be quantified as easily as a national residual load; the region to be supplied by a certain power plant fleet becomes increasingly dynamic. For the determination of the load schedule of dispatchable power plants, the national residual load becomes less important over time especially in countries with strong interconnections. When taking this into account, the potential market for baseload power plants shrinks far less quickly.

CCS CCGT plants have a lower utilisation than the other CCS technologies, but cover a wide range of applications; in the scenarios, CCGT plants typically start at a high utilisation in the early years, which decreases significantly over time. In some regions the decrease is extreme. In the first years of operation the utilisation is typically so high, that when looking at the LCOE only in these specific years, it would seem economic to replace such plant with a coal power plant. However, the lower utilisation towards the end of the scenario years would not justify the construction of a baseload power plant. In contrast, the open-cycle gas turbines are used only as peakers, running in a low number of hours. The decreasing utilisation of conventional capacities is not only the result of the increase in RES; it is also the consequence of the increase in EUA price and the decreasing CO_2 cap. The model has to keep utilisation of gas turbines low in order to keep emissions below the defined levels.

The FLH of wind, both onshore and offshore, decline in later years. The model uses the best potential steps first, while less attractive sites are used later. For PV, the opposite is the case: the politically defined targets of the NREAPs force the utilisation of PV sites that would otherwise not be utilised. In later years, the solver builds capacities only in Southern Europe, which increases average FLH.

For CSP, the decrease in the capacity factor due to the utilisation of less attractive sites is reinforced by another effect: In 2020, most of the heat captured by the collectors is used directly, i.e. without being stored. In later years, the diffusion of RES-E, especially PV, decreases the likelihood that the sunny hours are the most economic time to use the energy. With many PV systems installed, demand is often covered by wind and PV alone, meaning that the thermal energy of the CSP plants needs to be stored. This leads to losses, which decrease the FLH of the technology in the later years. In 2050, the largest part of generation potential has to be stored to be used.

6.1.3. Expansion of interconnection capacities

The shift towards renewable energies is enabled through the expansion of the transmission grid. More precisely, the model strengthens the interconnections between the countries, which is the only part of the grid covered in the model. Figure 6.5 shows the *Grid strength*, as defined in eqn. 6.2 as the cumulative product of length and transmission capacity of all interconnectors.

$$\text{Grid strength} = \sum_{\forall t} L_t^{\text{tot}} X_{ta} \qquad (6.2)$$

Grid strength, expressed in GW km, is highly correlated to material consumption and costs of the transmission grid. The value in 2020 is very close to the initial grid strength, which is approximately 29,000 GW km. This indicates, that the existing interconnection between the countries is almost sufficient for the needs in 2020. By 2030, the necessity to expand is moderate; despite an already high proportion of fluctuating RES-E. After 2040, grid expansion accelerates sharply; between 2040 and 2050, the model grid strength increases by 114 %.

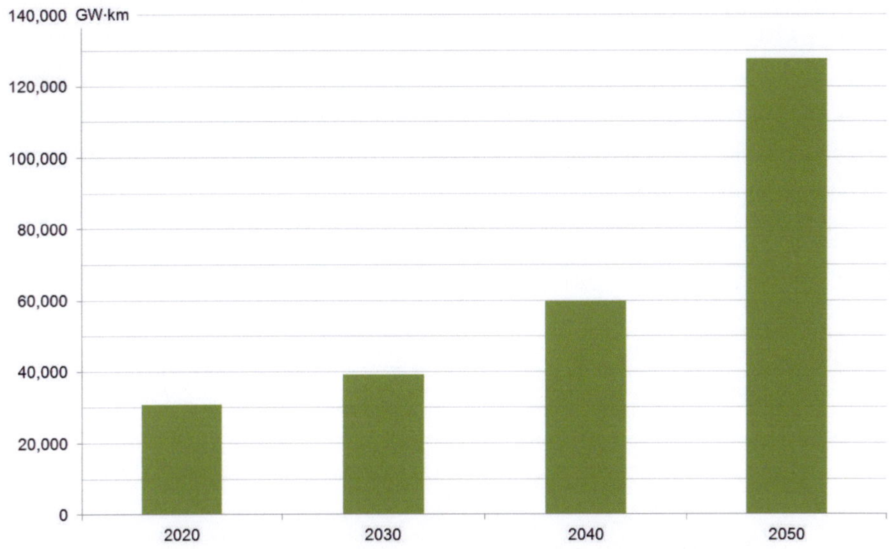

Figure 6.5.: Developments of grid strength in the OPT scenario.

The behaviour displayed is typical for optimising electricity system models; it is caused by the fact that interconnections are "cheap" compared to other power infrastructures. As it will be discussed in section 6.1.9, the part of the electricity grid that is referred to as interconnections makes up only a relatively small part of the total costs of the system. Increasing transmission capacity of a certain line can be beneficial in several ways: it can facilitate higher utilisations of dispatchable power plants with low variable costs; decrease the share of RE generation potential that has to be curtailed; or allow a more efficient usage of storage facilities. Facilitating these options though grid expansion becomes more necessary the more the system is "pushed to the edge" by extreme requirements, e.g. a very high enforced rate of decarbonisation. Under such conditions, grid enhancements are among the cheapest options the model can utilise and thus often the first to be implemented.

The final outlay of the interconnections is depicted in figure 6.6. Most of the interconnections have NTC between 3 and 6 GW. The strongest connection is built between France and Spain; this particular border is important for connecting the good wind and solar energy potentials of the Iberian Peninsula to Europe's mainland. Another strong connection is built between the UK and mainland Europe. France acts a central hub in Western Europe, connecting the British Isles and the Iberian Peninsula to the electricity storage facilities of the Alpine region; consequently, the country has the highest NTCs with its neighbours, totalling 47.6 GW. Germany, with a NTC of 42.6 GW with its neighbours is the second-best connected country and also functions as transit hub; it connects Eastern with Western Europe and the Alpine region, but is also connected well with Scandinavia. In Scandinavia, Norway is not only the biggest net exporter of power, but also acts as a source of dispatchable hydropower.

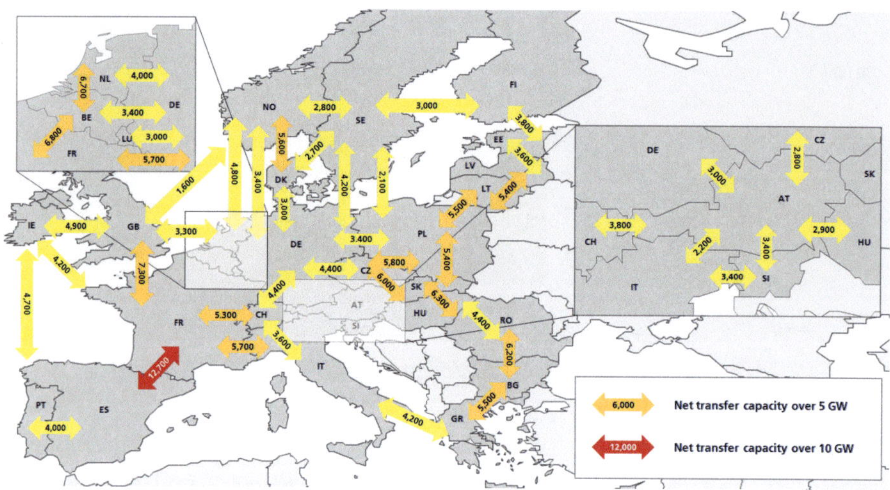

Figure 6.6.: NTC of interconnections [MW] in 2050 in the OPT scenario. Values above 5 GW are highlighted.

It should also be mentioned that the resulting grid outlay is very homogeneous. In earlier runs of the model, in which the expansion of RES-E did not take place endogenously, the model transformed certain lines into strong "backbones", often exceeding 20 GW for a single connection. The exogenously defined RE technology capacities were not very well balanced and thus needed a much stronger grid to match supply and demand. It seems that the homogeneous design of the grid is to some extent the result of the balanced distribution of RE technologies.

Nevertheless, it is obvious that the strong expansion of the grid would in reality be challenging due to issues that are not covered in the model: public resistance, conflicting interests between market participants with strong market power and administrative hurdles slow down the expansion or hinder the transmission capacities to reach the optimum values. Although the GRID scenario seeks to explore the impact of these issues by limiting the strength of the grid, optimisation models are not able to fully depict these problems.

6.1.4. Implications for the transmission and distribution grid

The results discussed above cover the interconnections in terms of net transfer capacity between countries; other components of the grid are not included. The copperplate-simplification implies that the countries do not have transmission bottlenecks within their own borders; the transport and distribution grids do not restrict the flow of electricity. The losses occurring within the borders are treated in the model as additional electricity demand. Consequently, the applied modelling method covers a significant share of the European grid, but assumes that the remaining parts are adequately developed..

This simplification is for two reasons not as restricting as it might seem: Firstly, the interconnection lines presented above are calculated between the weighted centres of demand. This means that they cover long distances and act as backbones connecting major demand centres; the lines would also be available for transport within the countries. The way the interconnections are modelled means that they cover a substantial part of the interior grids. Secondly, the transport lines within the countries would be congested mostly during times of high RES-E infeed, e.g. very windy hours. At these times, curtailment is already taking place in the model, since the interconnecting lines are not designed to transport all available RE supply at any time. This means that although congestion might occur on the internal transport lines, this is often superimposed by the curtailment induced by the congested interconnectors during these hours.

Modelling the remaining parts of the transport grid would require explicit modelling of each power line and a very detailed knowledge of the location of both load centres and generation capacities. This much more detailed spatial modelling is beyond the scope of this work and would be very challenging in terms of computational resources.

The lower voltage levels are completely excluded from the calculation. Although their inclusion is theoretically possible, the immensely increased calculating time would require simplifications in other aspects of the model. It is however clear that the changes in the structure of electricity generation towards decentralised generation require new grid concepts. The high growth rates of wind power and PV will become especially challenging in this context.

6.1.5. Electricity storage facilities

Although the development of the installed capacities of electricity storage facilities is included in figure 6.2, it is worthwhile examining it in detail (see figure 6.7). The 2020 values represent the storages existing today long with two PHES projects expected to be completed by 2020. As it can be seen, the storage capacity, expressed in peak turbine capacity, only increases by a moderate 14 %. The additional storage is built almost entirely in the UK. The results regarding electricity storage are somewhat counter-intuitive; they contradict the frequently expressed assumption that the expansion of RE technologies should be complemented by the expansion of storage facilities.

To understand the result it is important to stress that there is no technical necessity for storage technologies: even in a situation with high excess electricity from fluctuating RES and electricity demand a few hours later, it is possible to curtail the generation potential and supply the demand alternatively, e.g. with gas turbines. Consequently, there can only be an economic benefit, rather than a technical need for storage[5] but only economic benefit from it. The solver decides to build storage only when it is advantageous from an economic point of view. This means that the benefits of utilisation must outweigh the capital costs, which is the case only at a certain value of FLH, with the actual level depending on many factors. The solver usually prefers electricity exchange over storage utilisation, due to the lower losses, at least up to a certain transport distance.

This feature is exemplified in figure 6.8, which shows the power mix in the UK in calendar week (CW) 47 of 2050 in the UK.System load, domestic electricity generation and imports are depicted above the horizontal axis. Whenever generation plus imports exceed demand, one or more of the options below the x-axis is employed: either power is exported or the storage facilities are filled (e.g. pumping into PHES) or RES-E generation is curtailed. The difference between load and generation plus imports will be called *excess electricity*, in the following paragraphs.

---

[5]Theoretically, there are exceptions to this, for example if the power sector has to be decarbonised, *completely* and new nuclear power plants or negative emissions from biomass CCS are not an option.

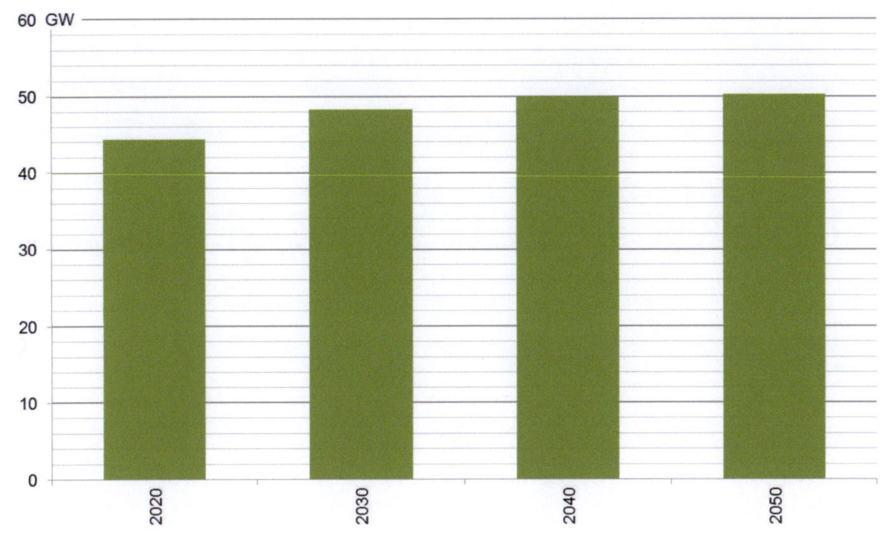

Figure 6.7.: Installed electricity storage facility capacities in the OPT scenario, expressed in peak turbine capacity.

The situation shown in figure 6.8 is an example, but is not "typical"; due to the high proportion of fluctuating RES-E in the system, the similarities between the weeks are small. This particular week is windy, though not exceptionally so. On Monday and Tuesday the CCGT power plants have to cover a significant portion of demand, supplemented by imports. Between Monday night and Tuesday morning, power from wind energy exceeds domestic demand for the first time in this particular week. This demonstrates a typical behaviour of the model: export is usually the first choice in utilising excess electricity[6]. The main reason for this is that the losses generated are usually significantly lower than those generated by storage. Consequently, storing electricity is usually only the second best option, as it essentially wastes electricity. It is only used if there is a significant difference between the (implicit) value of power in different hours. Please also note in figure 6.8 how during noon on Thursday the storages generate power for export, as domestic RES-E already meets demand; electricity storages are, like generation capacities, to some degree shared between countries.

When the installed infrastructure capacity is assumed to be fixed, curtailment

---

[6]It has to be borne in mind that during the model runs, excess electricity is not a constant: the model influences it by increasing or decreasing investments in RES-E capacities and by adjusting generation from dispatchable power plants.

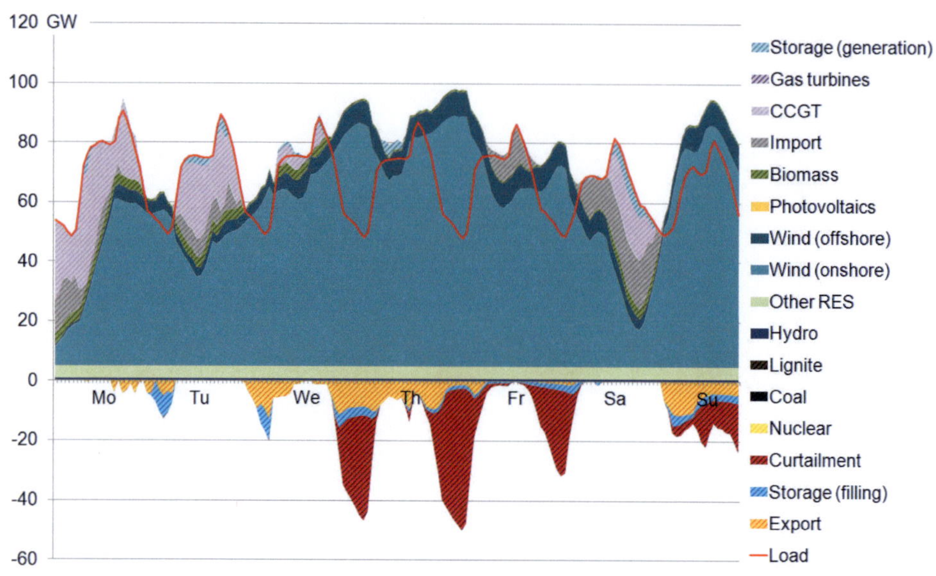

Figure 6.8.: Power mix in the UK in CW 47 of 2050 (OPT scenario).

is least-favoured option below the x-axis, in the model. It means that "free"[7] RE generation potential is wasted. However, in this sample week alone, the UK curtails over 960 GWh. Over the year, this totals 18.5 TWh, which is the highest value of national curtailment in the OPT scenario. However, the model concludes that utilising this energy is too costly; it would only be possible with storage and to some extent through increasing the interconnections. The storage facilities of 663 FLH are utilised in the UK in 2050, which is above the European average of 558 FLH for the same year. The low value shows that storage is only used if no better option exists.

6.1.6. Renewable electricity utilisation and curtailment

As already discussed, the scenario relies on a high share of RES-E for the decarbonisation of power supply, with the majority of RES-E coming from fluctuating sources. Under such circumstances, utilising the full generation potential of the installed capacity is usually not economic. The infrastructure that would be necessary to utilise the last MWh of a site is too expensive due to their low utilisation, regardless of whether storage facilities or grids are built. Consequently, a proportion of RES-E is curtailed, but as previously mentioned, the model does not decide

---

[7]Of course the energy is only free of cost for given, fixed RE capacities.

which technology. Figure 6.9 shows the share of curtailed power. *Gross curtailment* refers to the ratio of curtailment to total generation potential of all technologies. In turn, *net curtailment* refers to the ratio of curtailment to the generation potential of wind power and PV, because their fluctuations are to some extent the origin of the curtailment.

$$\text{Gross curtailment in year } a = \frac{\text{Total curtailment in year } a}{\text{Generation potential of all RE in year } a} \quad (6.3)$$

$$\text{Net curtailmentin year } a = \frac{\text{Total curtailment in year } a}{\text{Generation potential of wind and PV in year } a} \quad (6.4)$$

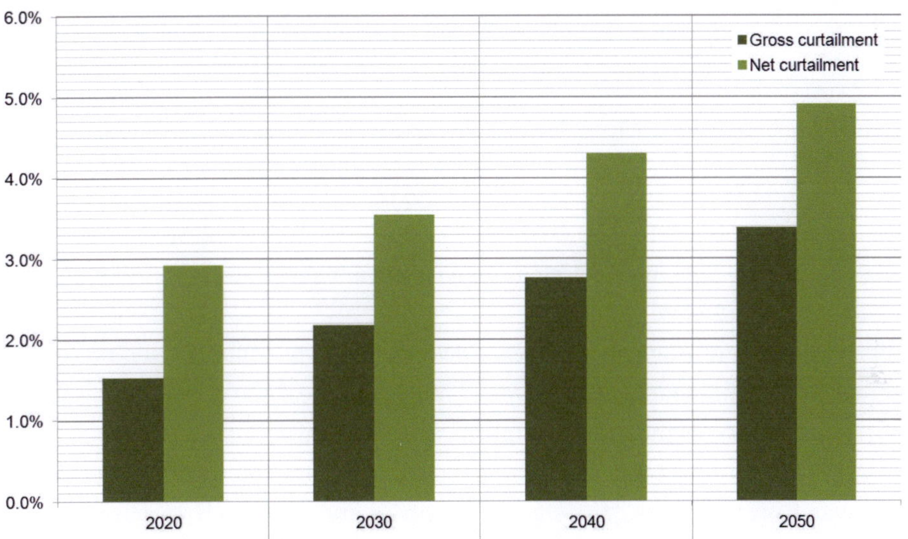

Figure 6.9.: Development of gross and net curtailment over time in the OPT scenario.

As it can be seen in figure 6.9, both net and gross curtailment increase considerably over time, with net curtailment reaching almost 5 % in 2050. These values are the result of the implicit assumption that grids within the countries are built so that no bottlenecks can occur (i.ethe copperplate-simplification that was already discussed above). In reality, the parts of the grid that are not covered by PowerACE-Europe are also subject to cost-benefits analysis, meaning that bottle-

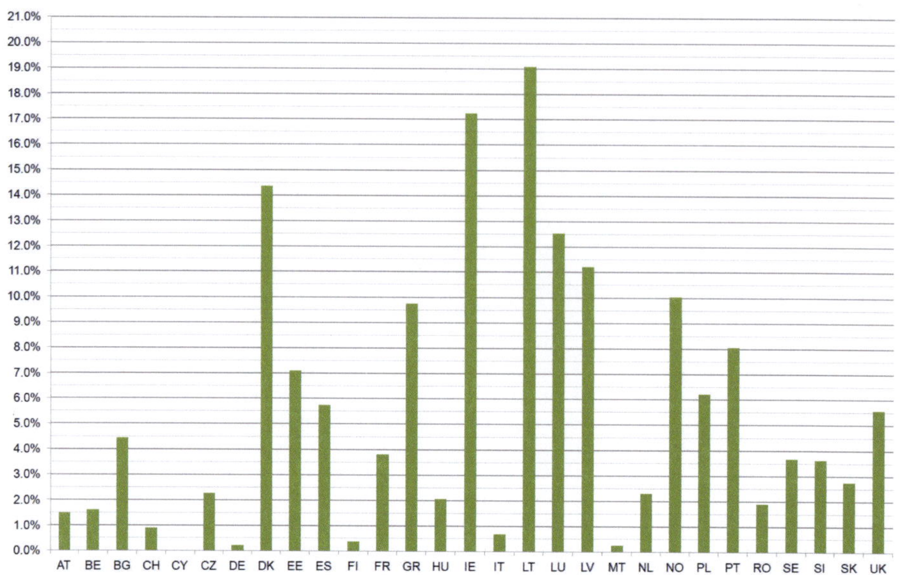

Figure 6.10.: National net curtailment in 2050 in the OPT scenario. Country names are here and in the following figures abbreviated by their ISO 3166-1 alpha-2 code.

necks will occur. This implies that in reality curtailment is likely to be even higher in a least-cost strategy.

Furthermore, curtailment varies significantly between countries, as indicated in figure 6.10. The individual values depend largely on the share of fluctuating RES-E, the connection to neighbouring countries and the availability of storage facilities. The NTC of a country to its neighbours is strongly correlated with its total electricity demand, which means that smaller countries are more likely to have a greater need for curtailment. More than 10 % of fluctuating RE generation potential is curtailed in Lithuania, Ireland, Denmark, Luxembourg and Latvia .

6.1.7. Electricity trade and national balances

The national import and export balances provide information useful for understanding the design of the power system constructed by the model. In general, exchange of electricity between countries increases over time, which can be monitored for example through the losses on the interconnectors. In the OPT scenario, endogenously calculated grid losses increase from 14.3 TWh to 42 TWh, i.e. an increase of 192 %. This demonstrates the need for cooperation between the countries that exchange electricity in increasing volumes and over longer distances.

Analysis of the losses do not generate conclusions on the netted imports and exports and the national trading balances. In the following, the *import dependency* of a country will be defined as follows:

$$\text{Import dependency} = \frac{\text{Net imports}}{\text{Net electricity demand} + \text{interior losses}} \tag{6.5}$$

The import dependencies are shown in table 6.1 for all scenario years.[8] That Estonia is the biggest net exporting country is a function of the denominator of the indicator; the low demand, in combination with good wind sites leads to net exports of 20 TWh. As with the (in some case considerable) changes within 10 years this is a peculiarity of the smaller countries and should not be over-interpreted.

The high exports of Norway demonstrate the function of the country in the power system. In current discussions, Norway's role in the mid- and long long-term future is often focussed on the country's storage possibilities. Tapping this potential does not necessarily require the construction of pumped storage hydropower plants. The high capacities of reservoir hydropower facilities allow a throttling of generation in times of excess electricity in the region. The saved hydropower generation could then be used when generation from fluctuation RES is low. That way, the combination of domestic demand and reservoir hydropower essentially behave like a very large electricity storage that can absorb electricity without additional pumps.[9] In the OPT scenario (and in the other scenarios as well), Norway's role is more that of dispatchable powerhouse. As it can be seen in figure 6.11, the model uses the flexibility of Norway's hydropower system to complement its wind power, thus tailoring the dispatch to meet the residual demand of the countries in the region. Please note that flexibility is limited to some extent by the fact that 5.3 GW of the installed hydropower capacities are run-of-river, for which a constant generation profile is assumed.Nevertheless, wind power increases in the model to levels at which even the flexibility of hydropower does not prevent curtailment. This can be seen during the evening and night of Thursday. In other weeks, curtailment is often much higher: Norway's net curtailment reaches 10 % in 2050. On Friday and Saturday, the country behaves like the storage option discussed above, though this happens only sporadically throughout the year.

---

[8]The national balances including net import and export can also be found in figure B.1 in Appendix B.

[9]Interestingly, with the previous version of PowerACE-Europe without endogenously calculated diffusion of RE technologies, exactly this happened in the scenarios.

Table 6.1.: Import dependencies and net imports of the countries in the OPT scenario, sorted in ascending order by their import dependency in 2050.

Country	Import dependency				Net imports [GWh]	
	2020	2030	2040	2050	2050	
EE	20%	5%	2%	-138%	-	20,211
NO	-28%	-38%	-59%	-91%	-	116,250
LV	-96%	-66%	-38%	-91%	-	12,198
IE	-13%	-4%	-6%	-70%	-	32,424
AT	-27%	-27%	-30%	-31%	-	22,944
DK	-68%	-42%	-24%	-22%	-	12,803
SI	21%	15%	20%	-16%	-	3,632
GR	-1%	-1%	3%	-13%	-	10,522
BG	-12%	-7%	-8%	-12%	-	6,775
SE	-3%	1%	-2%	-9%	-	14,287
CZ	18%	-9%	-12%	-6%	-	6,733
ES	-1%	0%	4%	-3%	-	12,876
PL	4%	1%	4%	-1%	-	2,779
CY	0%	0%	0%	0%		0
MT	0%	0%	0%	0%		0
FR	-16%	-6%	-2%	0%		1,405
RO	2%	0%	0%	1%		1,056
PT	7%	4%	-2%	1%		958
LT	-15%	30%	6%	2%		276
IT	2%	2%	3%	2%		10,872
UK	2%	0%	1%	5%		28,784
DE	6%	5%	2%	9%		57,901
NL	8%	7%	13%	11%		18,254
BE	22%	6%	-2%	16%		19,021
HU	15%	6%	6%	16%		10,801
SK	-1%	5%	8%	21%		15,989
LU	67%	34%	44%	30%		2,726
CH	38%	27%	25%	31%		31,164
FI	18%	16%	13%	32%		33,674

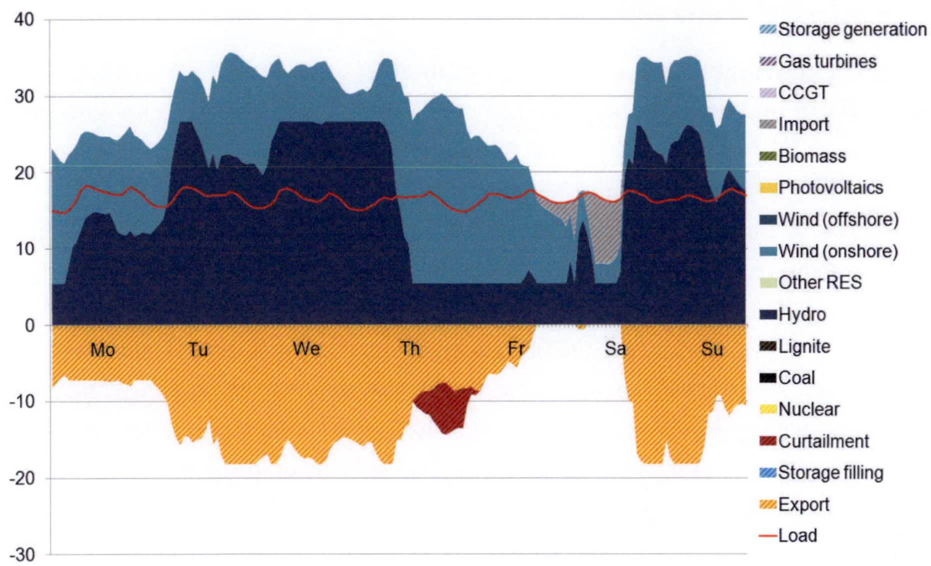

Figure 6.11.: Power mix in the Norway in CW 12 of 2050 (OPT scenario).

A similar abundance of reservoir hydropower, complemented with PHES systems, is also the driver behind Austria's export. In most other exporting countries, like Denmark and Ireland, the export is driven predominantly by their economically attractive RE potentials.

The opposite is true for the importing countries: although these countries may have favourable conditions on many sites, demand is too high to be met by these alone. Many of the importing countries have an above-average share of conventional generation in their supply-mix in 2050; the imports save fuel for the conventional plants, which are in turn also used as back-up capacities for neighbouring countries.

All import dependencies are below 32 % in 2050, with the highest values being limited to smaller countries. The import dependencies appear to be low given the large differences in attractiveness of RE potentials in Europe. It seems that the model designs a system in which most countries keep a high degree of self-sufficiency, despite the increasing necessity for exchange. Furthermore, a high import dependency does not necessarily mean that a country does not have the generation capacity to meet its own demand. The (political or economic) power due to electricity trade, that some countries have over others, is limited and in most cases is reciprocal.

6.1.8. CO_2 emissions and prices

Analysing the carbon emissions brings few insights, because they are limited by the emission cap and decrease to 75 Mt/a by 2050 in all scenarios.[10]. Total emissions decrease to 75 Mt/a in 2050 in all scenarios. What is more revealing than this development is the carbon price necessary for such a transformation.

PowerACE-Europe applies two direct ways to decrease emissions: the first one is the exogenous carbon price, which is assumed to be the price of EUAs in this study.[11] It is included in the costs of all processes with direct emissions, i.e. processes burning fossil fuels. The second method is the cap applied to the total annual emissions. For this constraint of the LP the *shadow price*, i.e. the value of the Lagrange multiplier of the optimal solution, is evaluated. "The shadow price on a particular constraint represents the change in the value of the objective function per unit increase in the righthand-side value of that constraint" (Bradley et al., 1977, p. 15). Consequently, shadow prices can be used to determine marginal CO_2 abatement costs.

In theory and in practice, the exogenous CO_2 price and the shadow price of the emission constraint have equal impact. The sum of the CO_2 price and the shadow price represents the fictitious carbon price level that would be necessary to keep emissions from the power sector below the cap. In the following, the sum of the two prices will be called the *effective carbon price*. It is a relatively good indicator for the pressure to decarbonise in the scenario and year. It should not be seen as a realistic carbon price, as it

a) only covers developments in the power sector,

b) assumes perfect foresight and a price-inelastic demand, and

c) assumes other support mechanisms are lacking, except the policies implemented to meet the NREAP targets and the exogenous RE technologies.

With this in mind it can, however,estimate the carbon price necessary, as the sole support instrument, to decrease emissions to the defined levels. The resulting values are shown in figure 6.12.

In 2020 the shadow price equals zero, as the cap is not effective. Through the exogenous CO_2 price of over 30 EUR/t emissions stay below the cap of 700 Mt. Although this is hardly visibly in the figure, the shadow price in 2030 is 1.1 EUR/t, meaning that the carbon cap does indeed have an effect; the exogenous CO_2 price

---

[10]The emission cap of 700 Mt in 2020 is, however, lightly undercut in the OPT scenario

[11]However, as international or intersectoral carbon trading is not included in the model, a carbon tax causing the same specific costs for emitting CO_2 would have the same effect.

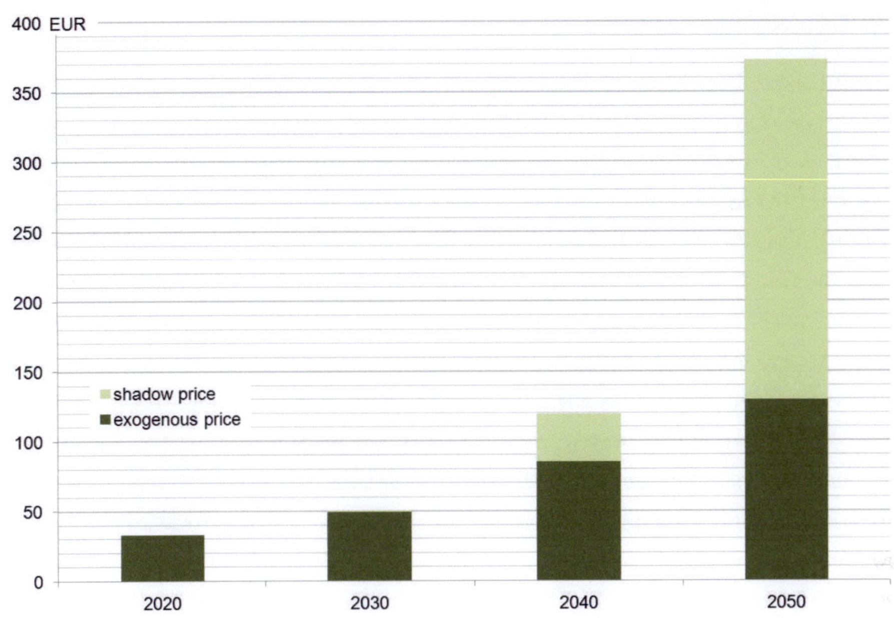

Figure 6.12.: Exogenous carbon price and the shadow price of the CO_2 cap in the OPT scenario.

alone would however result in only slightly higher emissions. In 2040, the shadow price of 34 EUR/t is significant and indicates the substantial intervention necessary to keep emissions below the cap of 150 Mt. The shadow price increases strongly in 2050, with the effective carbon price reaching 373 EUR/t.

Again, it is important not to over-interpret these figures. They indicate that by reducing the CO_2 cap by one ton, total system costs at this particular point would decrease by 373 EUR.[12] The cost reduction would be achieved through, for example, lower fuel consumption or slightly different investments in plants, interconnections and storage. The effective carbon price is influenced by the alternatives that are prevented by the cap; the availability of cheap yet carbon intensive generation options, e.g. lignite power plants, drives the price up. Furthermore, the figure does not allow conclusions to be drawn on the shape of the abatement-cost curve. It is possible that with a slightly less ambitious cap the effective carbon price would be significantly lower.

---

[12]More precisely, it would decrease by 373 EUR less the exogenous CO_2 price in 2050

6.1.9. Development of costs

The costs occurring in the OPT scenario are shown in figure 6.13 and the underlying data can also be found in table A.12 in Appendix A. Again, all monetary values are expressed in EUR_{2010} unless otherwise stated. Investments are annualised on the basis of overnight costs, i.e. capital cost during the construction phase are not taken into account. In 2020, the majority of costs are caused by conventional power generation and 41 % of the costs are attributed to conventional fuel costs. Up to this point, the newly constructed power plants are responsible for only a small proportion of costs, as the majority of plant already exists in 2010. For these plants, the fixed costs are estimated by values that are chosen to fit to the costs assumed for future plants.[13] Interconnectors and storage facilities are responsible for 1.9 % and 2.4 % of the annual costs, respectively. In this calculation, the specific costs of existing storage facilities are assumed to be identical to the future ones. In 2020 the costs of wind and solar power already surpasses the sum of all other RE technologies. The costs of the exogenous RE technologies are calculated by the model ResInvest. In this model, cost degression is partially endogenous through learning curves, which is described in detail by Held (2011). In 2020, 64 % of the cost of the exogenous RE technologies are attributed to hydropower, which in that year is also the exogenous RE technology with by far the highest installed capacities.

Until 2040 costs of conventional power generation, both fixed and variable components, remain at similar levels, though their share of total costs decreases steadily. Several developments almost cancel each other out: although base- and midload power plants become more expensive on average due to the diffusion of CCS, their installed capacities decrease. The decrease in power generation from fossil fuels, which would reduce total fuel costs, is counterbalanced by the rising coal and gas prices and the efficiency decrease caused by the utilisation of CCS technologies. Trivially, the costs for CO_2 transport and storage increase with the diffusion of CCS technologies but make up only 1.2 % of total costs in 2040.[14] Costs for the interconnectors increase at almost exactly the same rates as the grid strength. The only costs part that increases strongly in absolute figures is renewable energies. Cost of the exogenous RE technologies increase by 73 % by 2040, largely as a result of the expansion of biomass technologies. The costs of wind and solar technologies

---

[13]For nuclear power plants, costs are assumed to be slightly lower than for future reactors. For the other plant, costs are assumed to be similar but slightly higher than for 2020. Half of the gas power plants are assumed to be open-cycle and half combined-cycle power plants.

[14]As mentioned in chapter 4, this cost item is simplified to be variable costs, although in reality costs would largely consist of capital costs for the necessary infrastructure.

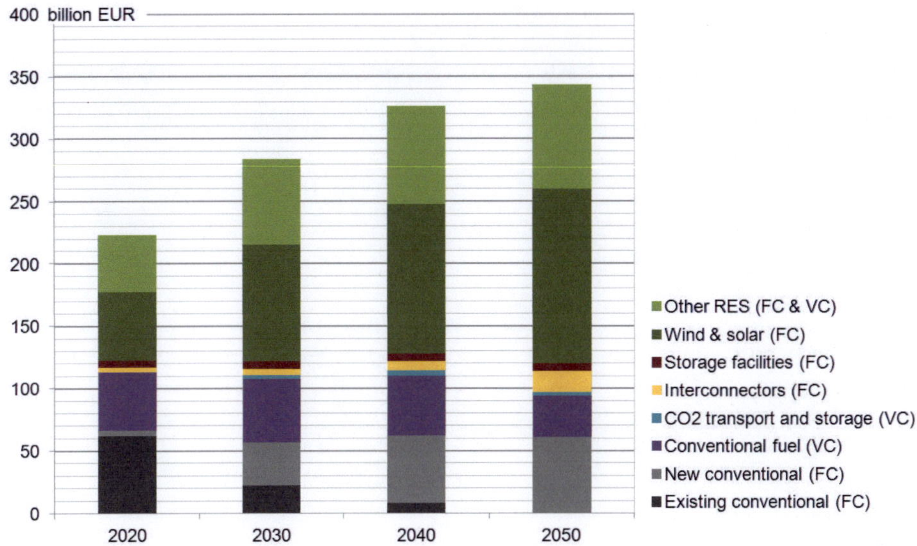

Figure 6.13.: Development of costs in the OPT scenario, including fixed costs (FC) and variable costs (VC).

increase by 118 %.

By 2050 it can be observed, that the model has a greater tendency to utilise several emission reduction options than in previous years. The interconnections are strong, using a higher proportion of the RE generation potential, and thus decreasing curtailment. The conventional power plants are utilised less, which reduces fuel expenditures. Consequently, conventional power generation is responsible for only 28 % of the costs in 2050. Interconnections account for 5 % and storage facilities for 1 %. RES-E accounts for 65 % of total costs covered by the model, and almost two thirds of these costs is attributed to wind and solar power generation.

As an optimisation model, PowerACE-Europe minimises all costs of the power system. The system boundaries and limitations have to be borne in mind when analysing the resulting costs as only those directly attributable to the electricity system are covered. Costs resulting from changes in the power sector, yet occurring in other sectors, are not taken into account. Such changes could occur for example if the heat supply from CHP plants changes.

In addition, not all costs in the power sector are covered. The biggest excluded elements are parts of the high voltage grid and the complete distribution grid. The former is partially taken into account through the connections between countries which cover parts of the internal grid. This is discussed in section 6.1.4. Nev-

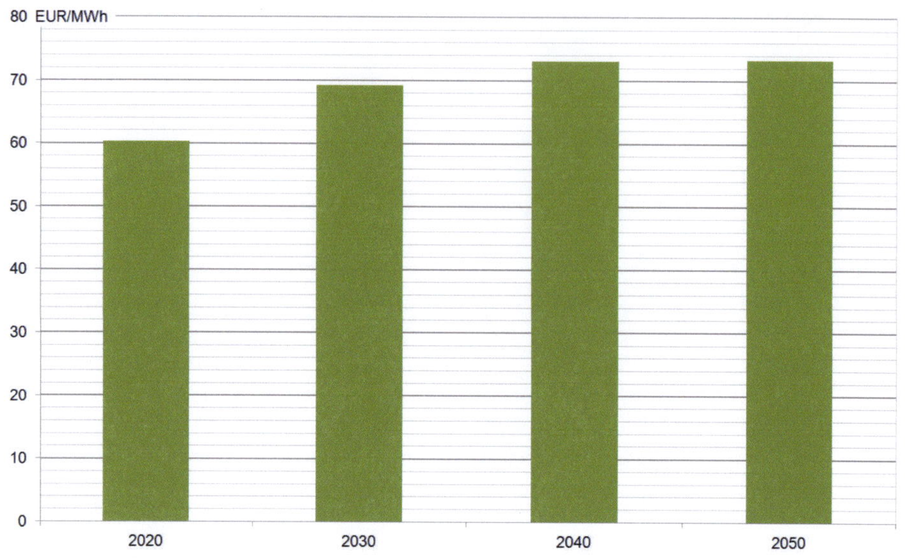

Figure 6.14.: Development of specific costs in the OPT scenario.

ertheless, realising a transformation process towards RE like the one depicted in the OPT scenario does not seem feasible without substantial reinforcement of the internal transport grid. Reinforcing the interior grids is, however, a prerequisite in all scenarios with a high proportion of fluctuating RES-E, which is the case for all decarbonisation scenarios. Therefore, the exclusion is assumed to not affect the conclusions drawn from the comparison between the scenarios.

Changes are also assumed to be necessary in the distribution grid, for example due to the diffusion of PV or smart grid applications. These changes affect the volumes and even the direction of energy flow in the lower voltage levels. Nevertheless, at least for the mid-term future, the majority of costs are expected to occur on the transmission grid, while costs for reinforcements of the distribution grid are predicted to remain moderate (Bundesnetzagentur, 2012, p. 27).

Furthermore, the expenditures for EUAs, though included in the optimisation process, are not considered costs in the discussion above. They are, in both their intention and impact, very similar to taxes, which are also not included.

Because not all costs components are captured and the market representation of PowerACE-Europe is rather simple, the model does not generate electricity prices. As discussed in chapter 3, making long-term prognoses about market prices is difficult because the future rules are almost impossible to predict. However, the model calculates specific costs of the generated electricity and supplied demand. In figure 6.14, specific costs (of electricity supplied) are depicted as:

$$\text{Specific costs of electricity} = \frac{\text{Total annual costs}}{\text{Net electricity demand} + \text{interior losses}} \quad (6.6)$$

It can be observed that specific costs increase throughout the whole scenario, with a steeper increase in the earlier years and a flattening towards the end. The increase of 22 % over the whole period seems moderate, considering the far-reaching changes in the system. Several points already discussed in this chapter cause this development. Relatively new and still expensive technologies are built, particularly at the beginning of the period; this concerns both RE and CCS technologies. In the later years these costs decrease through technological learning. This means, for example, that although the best wind sites are exploited early, cheaper and better wind turbines and power plants compensate for the inferior sites developed in later years.

6.2. Developments in the NoCCS, GRID and EFF scenario

The following sections describe the developments in the other scenarios, focusing on the deviations to the OPT scenario. The discussion follows the same structure as the previous sections.

6.2.1. Power generation mix and installed capacities

Figure 6.15 shows the development of the power generation mix and usage in all four scenarios, whereas figure 6.16 shows the installed generation capacities. The underlying data can be found in table A.2 to A.9 in Appendix A.

The challenge in the NoCCS scenario compared to the OPT scenario is to "replace" CCS plants with other technologies, while maintaining the emission level. As it can be seen, the model chooses a mix of nuclear power and gas. Surprisingly, the total generation from RE remains almost unchanged, although small shifts occur between countries. Power generation from onshore wind is even slightly lower in the NoCCS scenario. In turn, installed capacities of nuclear power are increased to 112 GW in 2050, which is almost today's level. As it can be seen in table 6.2,nuclear capacities are installed in almost every country that allow their construction, with France leading in the field.

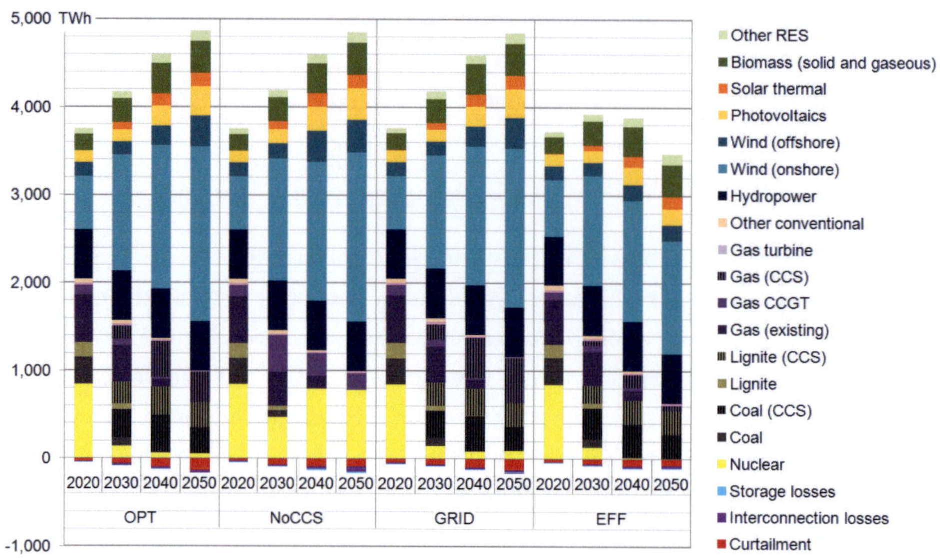

Figure 6.15.: Electricity generation, losses and curtailment in all scenarios.

Table 6.2.: Capacities of nuclear power plants in 2050 in the NoCCS scenario.

Country	Installed nuclear power [MW]
France	35,687
United Kingdom	14,571
Poland	13,549
Czech Republic	12,010
Romania	7,838
Slovenia	7,598
Slovakia	6,878
Hungary	6,370
Finland	5,977
Bulgaria	1,768

This result is very counter-intuitive, because the high proportion of fluctuating RES-E does not seem to be a good match for nuclear power. However, the model manages to maintain the utilisation of the units at 7,000 FLH in 2050, which is remarkably high given the circumstances. The way this is achieved is shown in figure 6.17. As it can be seen, the high NTCs between France are used to keep nuclear plants running at full power in most hours. In France, flexible biomass plays an important role in balancing demand and supply, while in other countries CCGT

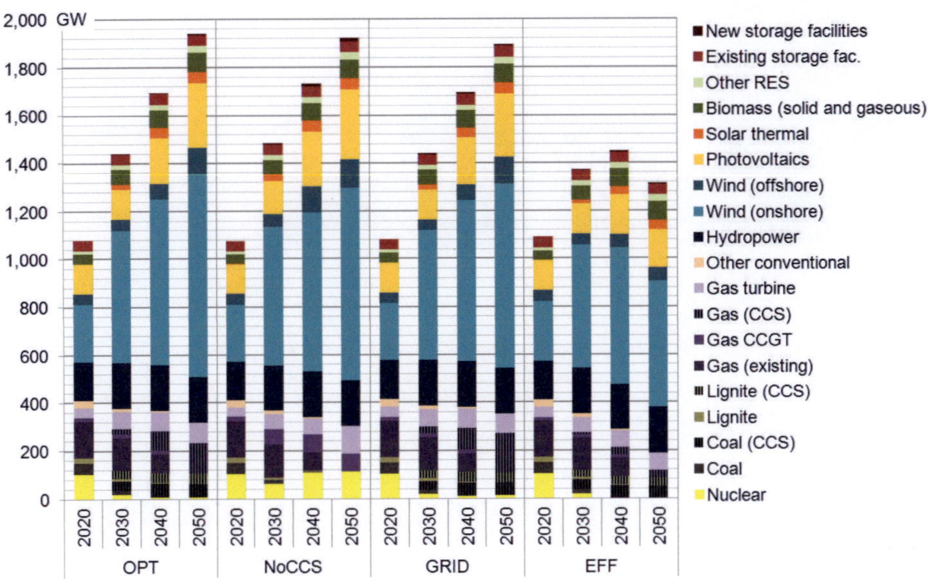

Figure 6.16.: Installed generation capacities in all scenarios.

performs this task. During Saturday and Sunday, nuclear power follows a very dynamic profile that seems challenging, if not technically impossible. Fast start-ups, followed by complete shut-downs of nuclear power plants only a few hours later, does not seem reasonable from today's point of view. However, it is possible that nuclear power plants will become more flexible over subsequent decades. Nevertheless, the results does not change significantly if nuclear power is modelled as being less flexible. In a sensitivity run, nuclear power generation was maintained above 60 % of the installed capacity, which is approximately the minimum generation for current nuclear plants. The changes to the overall results were insignificant, especially in terms of installed capacities, although curtailment was slightly increased. As the utilisation of the plant is already very high, the changes in system costs caused by a few additional hours of generation are low, especially due to the low variable costs of nuclear power.

In the NoCCS scenario, the 74 GW of installed capacities of CCGT power plants play an important role in the midload segment. Additionally, gas turbines capacities are 34 % higher than in the OPT scenario, but the utilisation of the plant is much higher. This is made possible by the carbon-free generation from nuclear power. A central conclusion is that the NoCCS scenario is characterised by a lack of technologies suited for the mid-load segment.

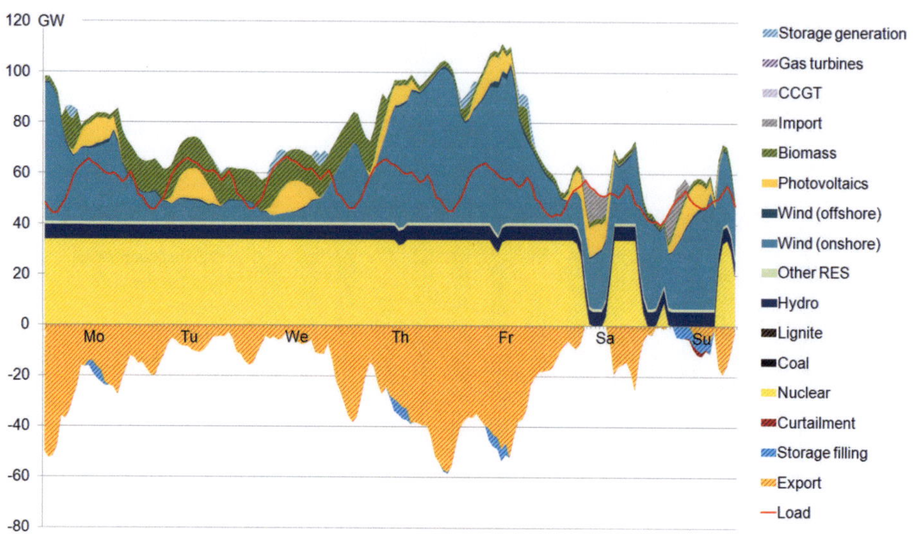

Figure 6.17.: Power mix in France in CW 32 of 2050 (NoCCS scenario).

The first observation regarding the GRID scenario is that it differs little to the OPT scenario in terms of installed capacities and power mix. The expected result of limiting the strengths of the interconnectors is a shift from renewable energies to conventional power generation. With fewer possibilities to use the full generation potential of RE sites, their economical attractiveness decreases. This behaviour can indeed be seen in the GRID scenario over the whole scenario horizon and the effect increases over time; however, the degree to which it takes place is relatively small. By 2050, 166 TWh are transferred from RES-E to conventional generation. The model relies less on onshore wind and PV, which decrease by 9.3 and 3.7 %, respectively.

In order to keep the emissions within the defined limits, the average emissions of conventional power must be reduced when increasing the market share. Therefore, in 2050 generation from coal and lignite CCS power plants is 9.9 and 4.1 % lower than in the OPT scenario. In turn, generation from gas and nuclear power increases by 45.9 and 65.8 %, respectively, but still remains very low in absolute figures. While power generation from coal, lignite, gas and nuclear energy in 2050 has an average carbon intensity of 75 kg_{CO_2}/MWh in the OPT scenario, it is decreased to 65 kg_{CO_2}/MWh in the GRID scenario.

The EFF scenario differs significantly from the OPT scenario in many ways. The lower demand leads to a substantially different power plant portfolio and generation mix.[15] The net RES-E share is consistently higher than in the OPT scenario and reaches 81.4 % in 2050. Due to the lower demand this also means that total generation from RE is 967 TWh below the OPT scenario. These differences are unevenly distributed among technologies: generation from offshore wind and PV is almost bisected, decreasing by 45.4 % and 48.1 %, respectively, whereas wind onshore decreases by 35 %. In 2050, the share of net fluctuating generation (i.e. subtracting curtailed power) on total power generation is 47 %, which is below the value of 54 % in the OPT scenario. This indicates that the power mix of the EFF scenario includes a higher proportion of dispatchable technologies.

On the conventional side of power generation, the model inclines towards the more carbon-intensive yet cheaper technologies. Power generation from coal and lignite in 2050 is only 5 % below the OPT scenario, which indicates a higher market share due to the significantly lower demand. In contrast, the contribution of CCS gas power plants decreases below 20 % of the value in the OPT scenario. This demonstrates that the high market share of gas power plants in the other scenarios is based mainly on its lower carbon content compared to coal and lignite. Generation from gas turbines more than doubles, although capacities are 18 % lower. Meeting the carbon cap is less "challenging"[16] in this scenario, which means that the gas turbines run more often instead of increasing capacities of cleaner but more capital-intensive technologies.

6.2.2. Utilisation and capacity factors of generation capacities

Analysing the utilisation of the power plant capacities in the scenarios reveals that the differences to OPT scenario are small; the figures will not be discussed here in detail. The baseload technologies are utilised very similar and show a comparable decrease over time. In the NoCCS scenario, CCGT plants replace their CCS pedants and are utilised similarly as these over the whole scenario horizon. The small differences in the FLH of RE technologies can be explained by higher or lower exploitation of the available sites; the FLH of all RE technologies except CSP deviate by +/- 6 % compared to the OPT scenario.

Significant differences can be observed for the utilisation of gas turbines; the developments over time are shown in figure 6.18. The technology is a important providers of dispatchability. Gas turbines are only built as peakload plants and have

---

[15] Although discussed later in detail, it should noted that the scenario underruns the emission cap sightly: CO_2 emissions decrease to 73 in 2050. Because the deviation is small, this is not considered an obstruction in comparison with the other scenarios.

[16] How "challenging" a scenario is for the model is for example expressed by the effective carbon price; that this indicator is not really precise is discussed later.

the by far lowest utilisation rates of all infrastructure in the system. Over time, their utilisation decreases, mostly because other technologies with lower costs push gas turbines out of the market. The increasing carbon price, both endogenous and exogenous, drives up the costs and thus worsens the position of the gas turbines.

In the NoCCS and and EFF scenario, the situation is different to that in the OPT or GRID scenario. In the NoCCS scenario, the closest alternative to open-cyle gas turbines are CCGTs (without CCS). In a power system without CCS, CCGT plant inclines towards the base-load segment and gas turbines tend towards the mid-load segment. In the EFF scenario, the lower effective carbon price has a similar effect of improving the relative position of gas turbines power plants in the (implicit) merit-order of technologies.

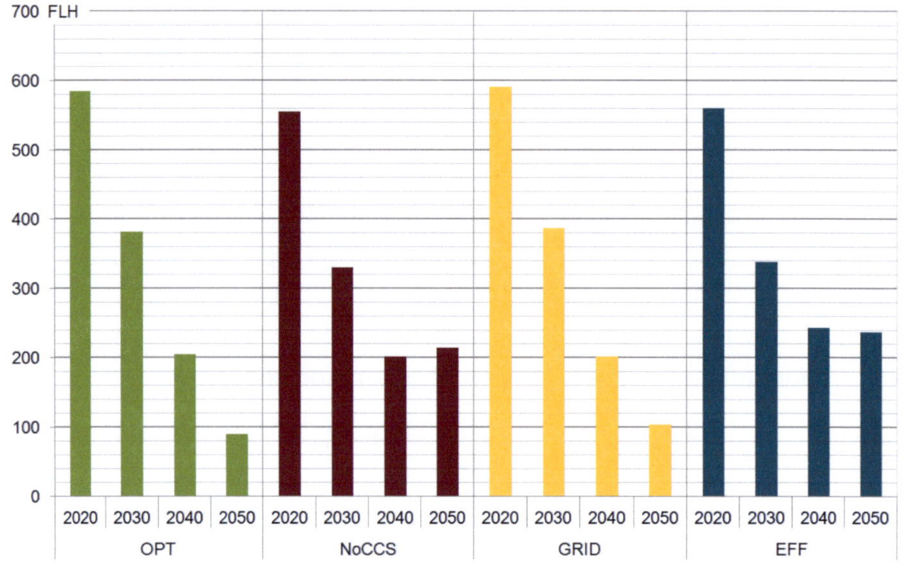

Figure 6.18.: Utilisation of open-cycle gas turbines over time in the scenarios.

6.2.3. Expansion of interconnection capacities

Significant differences exist within the grid development in the scenarios. The grid strength results are shown in figure 6.19. Until 2050, they increase between 182 % in the EFF scenario and 507 % in the NoCCS scenario. The NTC maps for 2050 can be found in figure B.2 to B.5 in Appendix B.

In 2050, grid strength expansion of the NoCCS scenario surpasses the already ambitious OPT scenario by 38 %. The rationale behind this has been touched upon in the previous section: The NoCCS scenario lacks ways to generate dispatchable power in line with the carbon cap. The value of carbon-free power is greater than that of the OPT scenario so therefore, its utilisation justifies higher costs. This is also visible in figure 6.15: the NoCCS scenario has a lower curtailment and higher grid losses because electricity is transported over longer distances. The high grid expansion is thus an expression of the challenges to meet demand if CCS is not available. As it can be seen in figure B.3 in the Appendix, the NTCs between the countries, as in the OPT scenario, increase rather homogeneously.

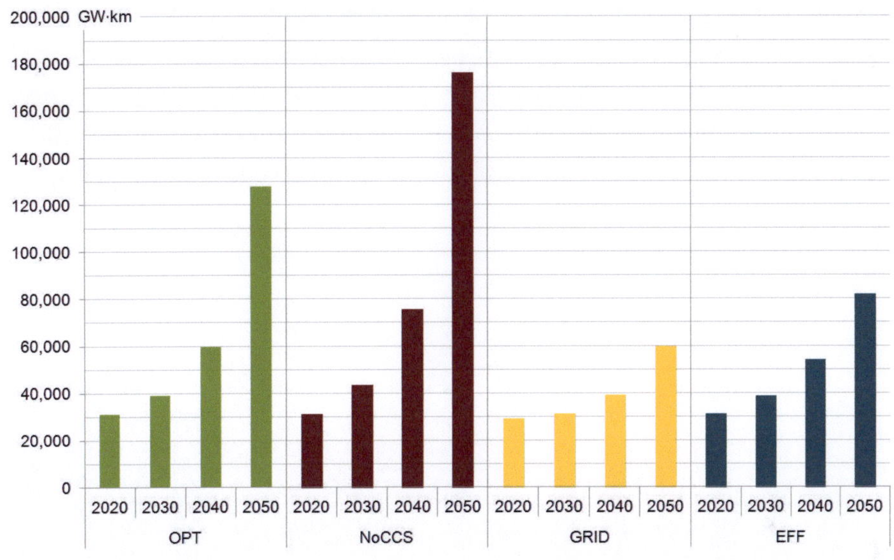

Figure 6.19.: Developments of grid strength in the scenarios.

In the GRID scenario the grid strength is input data, in contrast to the other scenarios. The value in each year a equals the value in year $a - 10$ in the OPT scenario. However, individual interconnections are not fixed; the model can decide to focus on important corridors. The NTC map in figure B.4 in the Appendix shows that only the connection between Spain and France has a NTC over 5 GW. Analyses comparing the scenario to the OPT did not reveal any particular patterns. It seems that the model maintains high connections to countries with storage facilities, though not exceptionally so. The general trend of a homogeneously strong grid also prevails in this scenario.

The grid strength in the EFF scenario increases by 182 % by 2050, which is significantly lower than in the OPT. Furthermore, the model does not show an extreme leap between 2041 and 2050, which is the case in the OPT and NoCCS scenario. Although the effect of a lower need for grid expansion could be expected, its size is surprising, as the optimal value for beyond today's expansion beyond levels is halved. The strong difference is likely to be a direct result of the lower demand, but also of the different supply mix with a smaller proportion of fluctuating or non-dispatchable RES-E. These co-benefits of successful energy efficiency measures should be taken into account in cost-benefit analyses.

6.2.4. Electricity storage facilities

The results of the other scenarios regarding electricity storage facilities are similar to the OPT results, which can be seen on the installed capacities depicted in figure 6.20. It has to be taken into account that the currently installed and planned capacities are already substantial and superimpose the results: In a situation without any existing storage, the differences between the scenarios are likely to be higher.

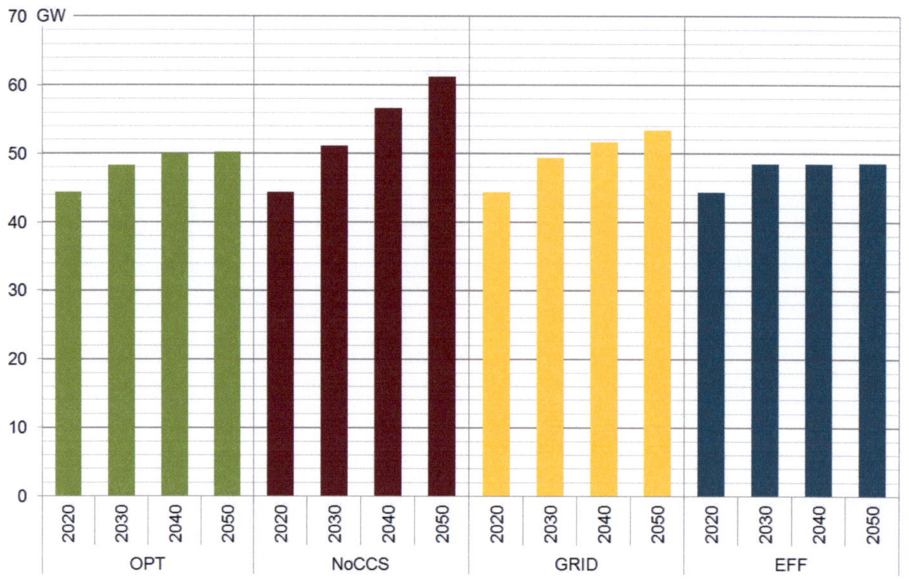

Figure 6.20.: Installed electricity storage facility capacities in all scenarios, expressed in peak turbine capacity.

The benefits of storage and the installed capacities are highest in the NoCCS scenario. In contrast to the OPT scenario, in which additional storage is only built in Britain, the conditions in the NoCCS scenario result in storage with turbine capacities above 500 MW additionally in Spain, Romania, Norway, and Denmark. Interestingly, these countries don not necessarily have exceptionally high proportions of undispatchable RES-E or a remote location in the grid, though these conditions are definitely drivers. Romania however has a relatively low proportion of wind and PV in the system (38 % before curtailment), but installs 1.1 GW of storage capacity. In this case nuclear power seems to be the main driver. To reach the high utilisation necessary to make nuclear power profitable despite the high proportion of non-dispatchable RES-E in the region, storage seem to be necessary.

The conditions for a storage systems scenario in the GRID are also slightly more favourable than in the OPT scenario. This was to be expected, but the differences are surprisingly small. The original expectation was that with restricted possibilities for building grids and balancing generation between countries, the model would either produce

a) substantially increased capacities of nuclear power or

b) substantially increased storage capacities.

Option a) would mean reducing the share of fluctuating RES-E in the system, thus reducing the need for transmission capacities. However, option a) is not cost-efficient, because under the given circumstances nuclear power is also dependent on a strong transmission grid. Option b) would indeed allow for a higher local utilisation of the fluctuating RES-E, but it is also not cost-efficient. What the model essentially needs to replace in the GRID scenario is a combination of dispatchability and "flexibility": The model faces difficulties in supplying demand during many hours that unproblematic in the OPT scenario, as the better grids allow spatially distant supply and demand to be matched. It seems that under the given circumstances, the most economic answer to this challenge is increasing conventional generation while reducing the average carbon intensity of fossil power by shifting towards the use of gas. This option is apparently less costly than the alternative of increasing storage capacities. However, it is subject to the availability of CCS: A combination of the challenges of the NoCCS and GRID scenario would most likely result either in significant investments into storage capacity or an increase of nuclear power, RES-E and curtailment.

The EFF scenario deviates only slightly from the OPT scenario, with significant storage systems built in the UK, Ireland and Denmark. The lower demand does not seem to substantially affect the need for storage in these countries. Additionally, the supply mix in these countries is quite similar in the two scenarios.

Overall, the conclusion that there is a low economic benefit of storage in all scenarios contrasts with the "need" for storage often presumed in non-scientific publications and discussions. Storage facilities are able to contribute to balancing the fluctuations of wind and solar power. However, under the conditions covered by the scenarios in this thesis, alternative options providing similar services are simply cheaper. Even where storage facilities exists, the model seeks to minimise their utilisation; the assumed losses of 20 % represent implicit variable costs making the usage of storage one of the least cost-effective options. In reality, losses will be higher for storage systems other than PHES.

The low economic benefit of new large-scale storage facilities appears to be a robust conclusion, considering that the techno-economic assumptions are optimistic. The properties of the storage dummies are chosen to resemble PHES, which will most likely remain the cheapest and most efficient large-scale electricity storage.

Nonetheless, smaller scale electric storage, for example batteries, could be beneficial for the distribution grids, which are not modelled in PowerACE-Europe. Furthermore, storage systems with additional use besides balancing demand and supply cannot be examined adequately by the model. Examples of such "dual-use" storages are the batteries of plug-in electric vehicles, or hydrogen that can be used in the electricity or transport sector. Regarding the latter, the "aversion" of the model towards (storage) losses and the low system efficiency of hydrogen render a large-scale diffusion in the model highly questionable. However, an unbiased comparison would have to include a detailed modelling of the transport sector; most internal combustion engines running on fossil fuel reach average efficiencies of only about 20 %, which is even lower than the system efficiency of hydrogen as a car fuel.[17]

6.2.5. Renewable electricity utilisation and curtailment

The developments of the net curtailment, i.e. the ratio of curtailment to the generation potential of wind power and PV, are depicted in figure 6.21. Evidently, the indicator's steady increase is a peculiarity of the OPT scenario. However, in all scenarios and years net curtailment is within a range of 2.8 and 5.1 %.

The curtailment which occurs is difficult to interpret, because it is influenced by many developments in the power system. In general, increasing generation from fluctuating RES or any other non-dispatchable technology can increase curtailment. Curtailment can be decreased, but this is no end in itself and is subject to the implicit cost-benefit comparison. This can be seen in the NoCCS scenario, in which

---

[17]By contrast, including the heating sector, for example heat pumps, or demand response would reduce the value of large scale storage facilities.

154

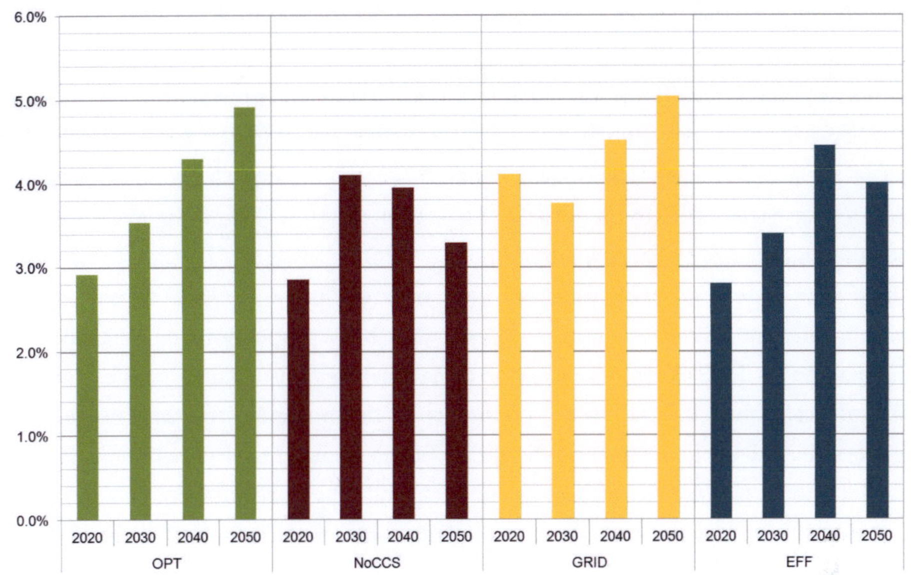

Figure 6.21.: Development of net curtailment over time in all scenarios.

curtailment steadily decreases after a strong increase between 2021 and 2030. As already discussed, this particular scenario is characterised by challenges to provide midload power; consequently, the model builds more grids and storage than in the other scenarios. Moreover, using these infrastructures becomes more competitive because the alternatives (e.g. curtailment and substitution with conventional generation) become more costly through the increasing effective carbon price. In other words, when confronted with a scarcity of low-carbon generation created by the absence of CCS, the model reduces the "wasting" of RE generation potential. A comparison between figure 6.21 and 6.22 indicates that curtailment is negatively linked to the effective carbon price.

In the GRID scenario, the minimum net curtailment occurs in 2030. This originates from the scenario's restriction that no grids can be constructed until 2021. The strong increase in RES-E up until 2020 is defined by the NREAPs and the model is restricted from building the grids that would be needed. This exemplifies the importance of coordination between grid development and RES-E. Drawing the conclusion that a slow development of the transmission grid is not critical, is untenable. The optimiser uses its perfect foresight and flexibility that it has to elude the limitations as well as possible. The system components are still well aligned to each other, which would probably not be the case in reality. In contrast to this, the year 2020 of the GRID scenario shows that without this flexibility, significant issues can occur.

The EFF scenario's development regarding curtailment are relatively unremarkable because it is very similar to the OPT scenario. It decreases in 2050 when almost no additional wind and solar capacities are built due to the lower demand, but transfer capacities are increased. What is interesting is that the lower demand does not significantly decrease curtailment.

It can be concluded that a certain level of curtailment is characteristic for the cost-efficient operation of a power system with high proportions of fluctuating or undispatchable RES-E. In the scenarios analysed net curtailment ranges between 2.8 and 5.1 %, but for individual countries or technologies, the values are much higher. For real-world systems it should be borne in mind that curtailment is, from a welfare point of view, in the interest of the public[18] and results from the interplay of many different system components. The costs of curtailing RES-E potential should be treated accordingly.

6.2.6. Electricity trade and national balances

Table A.13 to A.15 in Appendix A show the import dependencies of the countries over time for the NoCCS, GRID and EFF scenario, analogous to table 6.1 for the OPT scenario. The general trend can be compared with the standard deviation of the national import dependencies, which are shown in table 6.3.[19] As it can be seen, the values are significantly more dispersed in the NoCCS scenario and closer to the average in the GRID scenario.

Table 6.3.: Standard deviation of the import dependency in 2050.

Scenario	Standard deviation
OPT	0.3928
NoCCS	0.6836
GRID	0.1870
EFF	0.3135

As already discussed, more power is exchanged in the NoCCS scenario than in the other cases. In this context, nuclear power seems to play a central role. For example, France is the second highest exporter behind Norway and is not such

---

[18]Creating a power system without curtailment is possible but comes at significantly higher costs.
[19]The standard deviation is not completely unbiased in this context, as it gives the import dependency of countries with a low demand the same weight as countries with as high consumption. It should only serve as a guide.

an important net exporter of power in any other scenario. Despite this, no strong correlation exists between the nuclear capacities and the import-export balance. For example the UK is an importing nation in 2050 in this scenario, despite its nuclear capacities. However, electricity trade in general increases strongly in this scenario. This can be seen for example in the losses of the interconnections, which are 48 % higher than in the OPT scenario. One interpretation is that the lower variable costs of nuclear power make longer transport distances economically attractive. Still, the exclusion of nuclear power in some countries might also play a role.

The values of the import dependencies also reveal that not having CCS as an option forces some countries to reduce their self supply rate, while others become stronger exporters of electricity. This particularly affects countries that are either small or do not possess enough favourable RE sites and additionally have a nuclear phase-out policy. Their import dependency tends to increase. For example, the import dependency of Germany in 2050 changes from 9 % to 29 % between the OPT and NoCCS scenario. This finding adds up to the evidence suggesting that not having CCS as an option while having a relatively high electricity demand pushes the system to its extremes.

The opposite situation takes place in the GRID scenario; with significantly weaker transmission capacities, the countries are bound to draw back to domestic or at least regional supply options. This also means that the comparative advantage of favourable RE sites or complementary conditions, like Norway's storage hydropower system, cannot be exploited to its fullest. For example, Norway's exports in 2050 decrease from 116 TWh in the OPT scenario to 76 TWh in the GRID scenario.

The EFF scenario is very similar to the OPT scenario and the general trends remain the same. Nevertheless, the balances of some countries changes significantly. Switzerland, for example, is a net importing county in the OPT scenario, but becomes a strong exporter in the EFF case. Whereas Spain is a net importer in the EFF scenario, in contrast to all the other scenarios. Whether this is a robust result in the case of lower demand or an artifact of the shape of the demand developments over time can only be assessed through additional scenario runs and analyses.

6.2.7. CO_2 emissions and prices

Although all scenarios use the same exogenous input prices, they result in significantly different shadow prices of the carbon cap. The developments of both values are shown in figure 6.22.

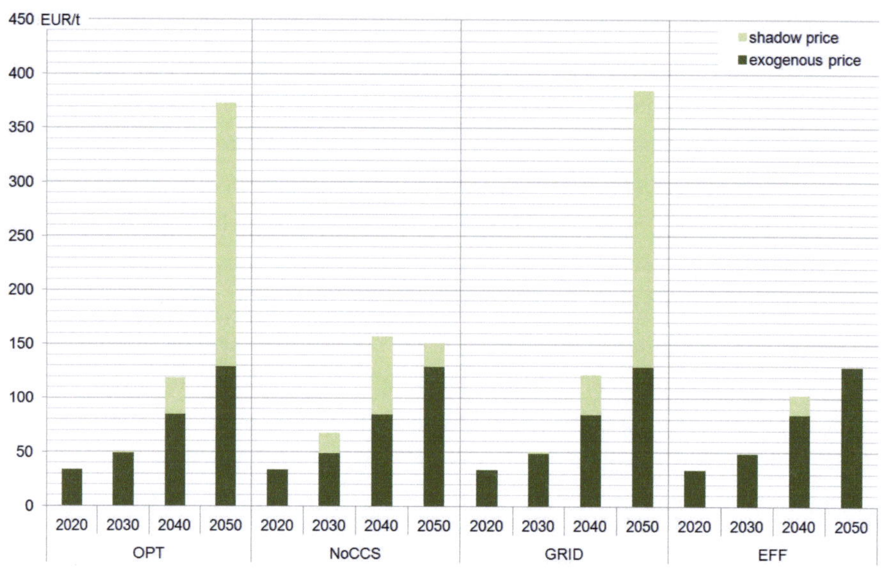

Figure 6.22.: Exogenous carbon price and the shadow price of the CO_2 cap in all scenarios.

In the NoCCS scenario the first peculiarity is that the effective carbon price is higher than in the OPT scenario in all years but the last, where it is less than half and even below the value for 2040. One possible interpretation is that the effective carbon price is linked to the opportunity costs of the emission cap, which again depends very much on the available alternatives. In 2040 in the NoCCS scenario, 91 GW of old capacities still exist. Although most of them are gas power plants with low specific emissions, their utilisation must be kept low in order to meet the carbon cap. Instead of using these existing and written off plants, the model must build new power plants with lower emissions. If the cap was less strict the existing capacities could be used. Consequently the opportunity costs at this particular point are the full costs including capital costs of the last unit built due to the emission cap less the *variable costs* of the cheapest plant is pushed out of the market through the emission cap. Without the carbon limitation, CCGT power plants would compare favourably. Therefore, the decreased utilisation is a result of the high effective carbon price that is observed for this year.[20]

The lower price in 2050 is the other side of the same coin, i.e. it is caused by the lack of alternatives. Without CCS, the next best option prevented by the cap is

---

[20]To illustrate this: The effective carbon price must be so high that the existing gas power plants are used very seldom while instead new nuclear power plants are built. It is understandable that this takes place only at a very high price.

158

closer to the one which is forced into the system by the cap, i.e. the opportunity costs are lower. However, the shape of the effective carbon price as a function of the emission cap is largely uncertain. Even monotony is not automatically ensured; this is the case only if the total system costs as a function of the carbon cap is convex. Although this appears to be the case for the scenarios presented here, cases deviating from this can be constructed. Theoretically, a shadow price of zero is possible if the mitigation-cost curve contains a flat plateau. Additionally, the intertemporal dependencies between the total carbon prices of the individual scenario years are not clear: The price in 2050 might also be lowered by the high price by 2040. Since most of the nuclear plants already have to be built for 2040, a lower cap in 2050 might have a smaller effect on total system costs.

An alternative or additional interpretation of the developments is that the chosen decarbonisation path is too fast for a scenario without CCS; a less strict emission cap may result in a smoother transformation process. This would mean that the achievable transition speed depends on the available generation technologies, which seems very plausible.

The GRID scenario is, as in most aspects, very similar to the OPT case, with effective carbon prices being 3.4 % higher in 2050. The limitations of the grid could be expected to drive up the effective carbon price, but the degree to which this takes place is again surprisingly small. The additional benefit of the higher NTCs between the OPT and GRID scenario seem to be positive, but small. Therefore, the alternative options for keeping emissions below the cap are only slightly more expensive.

A price decreasing effect was also expected for the EFF scenario, but in this case the effect is surprisingly strong: In 2040 and 2050, effective carbon price is 16.8 and 65.3 % below the values in the OPT scenario, respectively. This reveals that although the power system of the EFF scenario is in many aspects similar to the one of the OPT scenario, for example in the generation mix and proportion of RES-E, less "pressure" has to be applied to enforce the low emissions. This can be seen, for example, in the CCS gas power plants, which play a central role in the OPT and GRID scenario while generating only 70 TWh in 2050 of the EFF scenario. The technology is not competitive despite its favourable techno-economic assumptions and is only part of the solution due to the carbon cap. Although, as discussed above, the effective carbon price is not to be mistaken for an EUA price, it gives an indication of the effort needed to decarbonise a system. With the lower demand, fewer external incentives are necessary.

6.2.8. Development of costs

The development of the total costs are depicted in figure 6.23, and the underlying data can be found in table A.12 in Appendix A. The first striking observation is that the costs of the NoCCS and the GRID scenario are almost equivalent to the costs of the OPT scenario. Costs of the NoCCS in 2020 are almost identical to the OPT scenario but the difference grows over time and reaches the maximum of 14.0 billion EUR/a in 2040. After this, the difference decreases slightly to 13.1 billion EUR/a in 2050. This means that not having CCS as a decarbonisation option in the power sector increases the covered annual system costs by 3.8 % in 2050. The largest proportion of the additional costs comes from the high capital costs of nuclear power and the additional grid expansions. Fixed costs of conventional power plants and interconnections are 33 and 38 % higher in 2050, respectively. In turn, costs of fuels are 38 % lower.

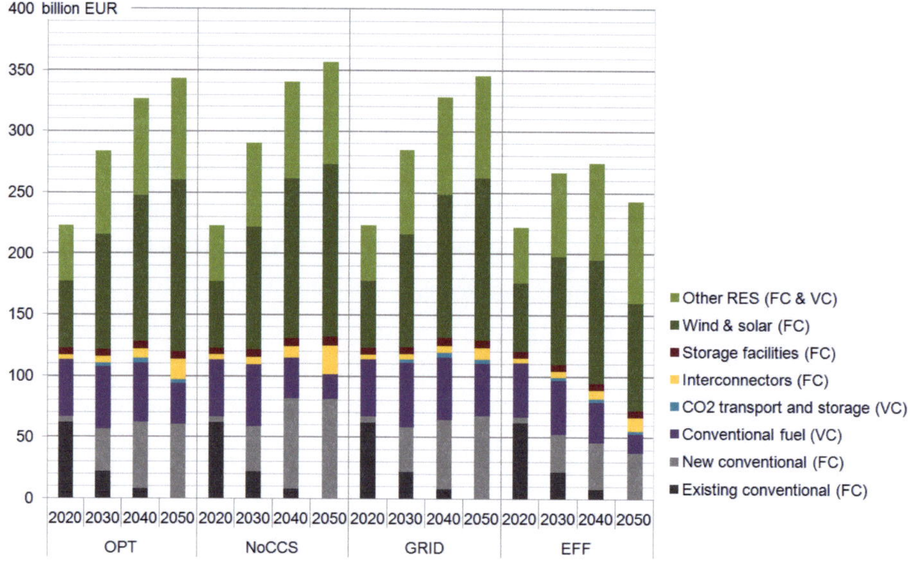

Figure 6.23.: Development of costs in all scenarios, including fixed costs (FC) and variable costs (VC).

In the GRID scenario, costs in 2050 exceed those of the OPT scenario by 2 billion EUR, i.e. 0.27 %. The cost of the interconnections decreases by 45 % and the costs of wind and solar power by 5 %. This is slightly overcompensated for by the increase in fixed costs of conventional power plants (+12 %) and fuel expenditures

(+29 %) as well as costs of storage facilities (+6 %). Although the increase in fuel consumptions, which originates from the shift from coal and lignite to natural gas, is especially substantial, total system costs are hardly affected. This reflects the surprisingly small impact that the limitation of the grid strength has on the system. This was already observed in many indicators. It endorses the evidence that grids are a relatively "cheap" part of the system. he grid enforcements allowed in the OPT scenario but excluded in the GRID scenario have a positive marginal utility. However, this utility seems to be small. The lower grid strength increases the system costs (on average) by 43,500 EUR per GW km of interconnection capacity not built in the GRID scenario. Another way to put it is that not building a 100 km interconnection of one GW transfer capacity saves annual grid costs of 10.2 million EUR per year, but causes 14.6 million EUR of additional costs in other system components. Naturally, the costs of the EFF scenario are significantly lower because demand is 28 % lower. Therefore, comparing system costs to the OPT scenario is less revealing than the changes in the proportions of these costs.[21] While conventional fuel costs decrease by 53 %, the CO_2 transport and storage cost decrease by only 20 %. Costs are reduced in the model by transferring from gas to coal and lignite, which, due to the assumed carbon capture rates, is only possible through the lower demand. Costs for wind and solar power decrease by 37.5 %, with generation from these technologies decreasing by 36.1 %.

Figure 6.24 shows the development of the specific costs per MWh. Again, these should not be confused with electricity prices and cover only the cost components included in the model. A comparison between the three scenarios with high electricity demand brings few additional insights, as the differences in costs have already been explained. However, the low differences between the OPT and the EFF scenario are striking: In 2050, specific costs are only 1.5 % lower. There are several reasons for this, the most important one being the lower degree of freedom regarding the RE portfolio. The generation from all RES, besides wind and solar power, is kept at the same level in all scenarios. These technologies have average costs of 79.8 EUR/MWh in 2050, which is higher than the costs of wind and solar power; average costs of wind and solar technologies are at 49.7 EUR/MWh in the OPT scenario and at 48.9 in the EFF scenario. This means that the RES-E mix of the EFF scenario is significantly more expensive than that of the other scenarios. Therefore, specific costs are not fully comparable. If the RES-E generation of the EFF scenario had

---

[21]It has to be borne in mind that some costs, most notably regarding the exogenous RES-E, are not controlled by PowerACE-Europe and are equal in all scenarios.

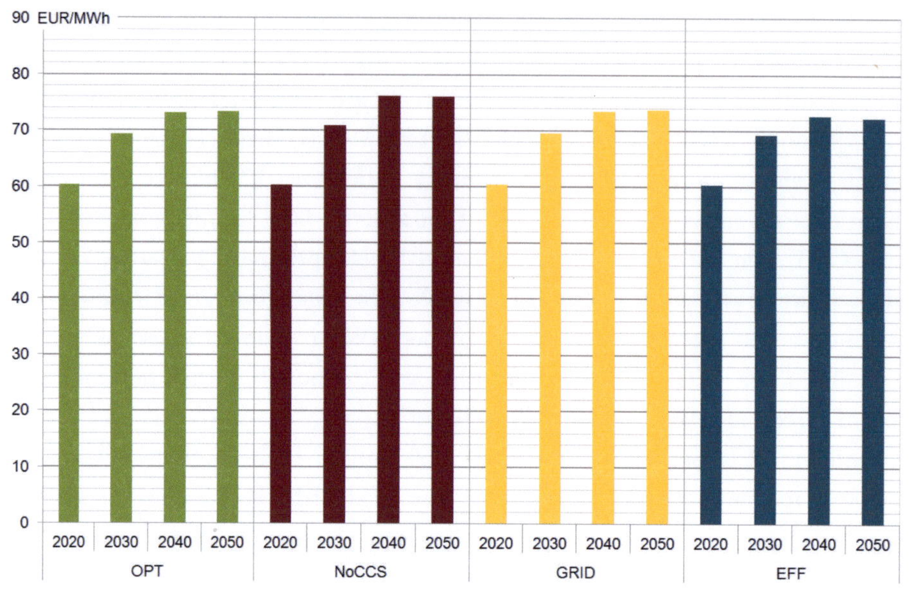

Figure 6.24.: Development of specific costs in all scenarios.

the same specific costs as the OPT scenario[22] the result would be different: In this case, specific costs in 2050 decrease to 70.1 EUR/MWh, i.e. 4.3 % below specific costs of the OPT scenario.

Trivially, the absolute costs for decarbonising the power supply depend largely on the electricity demand. The results of the scenarios suggest that the relative differences in costs of the different pathways with the same energy demand are only moderate, although change in absolute figures can reach 14.0 billion per year. Nevertheless, total costs of the modelled components can only be reduced substantially by reducing demand, which is in turn likely to generate costs.

Whether or not this equability is a robust result or a peculiarity of the modelling approach is difficult to assess. On the one hand, the model, like all optimisation models, will employ available options up to the point where they all have the same or similar marginal utilities: In every solution, changing the value of one variable that is not limited by a particular boundary, will result in a very small increase in the objective function. This means that the costs react rather insensitively to changes in most input parameters. Nevertheless, it cannot be disputed that solutions were

---

[22]This could be realised by reducing the generation from the non-hydro exogenous technologies, especially biomass.

calculated by the model for the NoCCS and GRID scenario with costs relatively close to the unrestricted OPT conditions, despite substantially more difficult conditions.

6.3. Sensitivity analyses for selected parameters

In order to understand how the model reacts to changes in certain central assumptions, several sensitivity analyses are performed. However, because each model run takes approximately 26 to 36 hours, depending on the individual settings, the number of runs is restricted.

In model-based electricity system scenario analyses a typical sensitivity analysis is testing different fuel price scenarios. This is performed because the development of these parameters has a high impact, especially on the short- and medium-term horizon. Furthermore, the insecurities regarding the future prices are high. However, for this study, other parameters seem much more crucial. The fuel price developments are superimposed to a large degree by the developments of the effective carbon price. The relationship between the price levels naturally does matter, but due to the difference in carbon content of the fuels, the impact of deviating prices on installed capacities and utilisation is strongly dampened.

Instead, sensitivity analyses are performed for parameters for which the impacts cannot be easily predicted i.e. the interest rate used for discounting, the decarbonisation rate and the specific investments of key technologies.

6.3.1. Interest rate

Testing the impact of the discount interest rate i^d is highly relevant: The impacts are difficult to assess due to the complexity, and as discussed in section 4.5.3; no consensus exists on appropriate values. To exemplify the general trends, the OPT scenario is calculated again with 2 % and 10 % discount interest rate, nd the results compared to those of the default value of 6 %. Figures 6.25 and 6.26 show the generation and installed capacities in the respective scenarios.

The impact of i^d is heterogeneous; in general, a high discount rate reduces the weight of costs occurring in later years of the optimisation problem. Consequently the higher net RES-E share in 2020 when applying a 10 % rate is surprising. It has to be seen in the context of the changes in the long-term strategy; the model reduces the construction of the gas turbines in 2020 by investing earlier into onshore wind turbines. The generation mix in 2050 is relatively stable, i.e. the influence of i^d is small. Due to the carbon cap, the optimal power mix depends more on the specific emissions of the technologies than on the costs differences due to i^d. However, differences exist in the installed capacities of conventional power plants. At higher i^{d}, the model builds cheaper yet more carbon intensive CCS coal and lignite power plants in 2030 and 2040. For example, the capacities of these two technologies installed in 2030 are 93 % higher at $i^{\mathrm{d}} = 10$ % than at $i^{\mathrm{d}} = 2$ %. In the earlier years, this reduces costs, but increases costs in later years, when there is low utilisation of these power plants. At an i^{d} of 10 %, the late disadvantages have low weight. The higher expenditures for RES-E in the beginning are thus the result of a conventional power plant park optimised for low costs in the short run. A pre-drawing of RE investments is, under the given assumptions, superior to investments into expensive gas power plants.

In conclusion, the effect of the interest rate applied in discounting has a significant effect especially on investments in the first two decades. The restrictions regarding emissions dominate developments in the later years. All conclusions drawn from the scenario results presented in this chapter are seen as robust to changes in the applied discount interest rate.

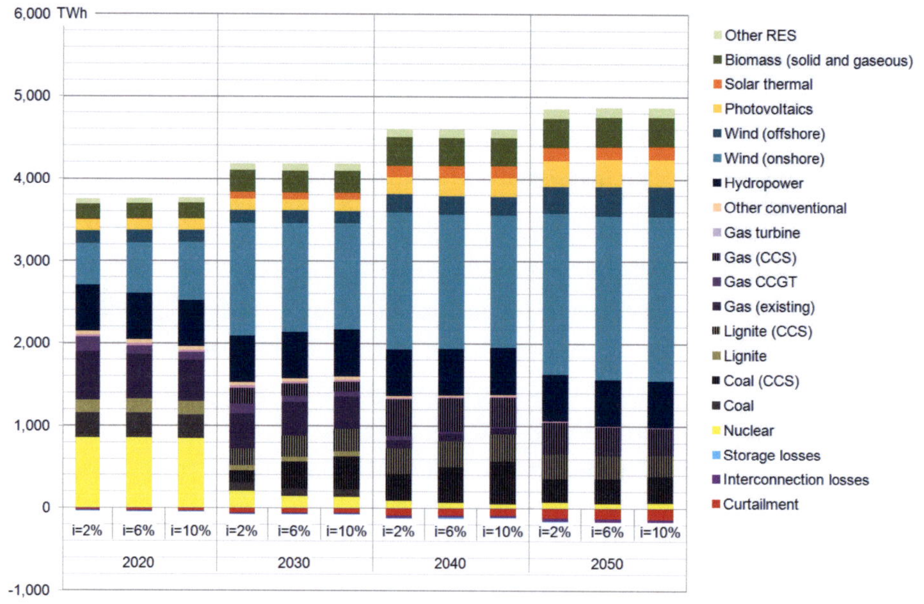

Figure 6.25.: Electricity generation, losses and curtailment for different discount interest rates.

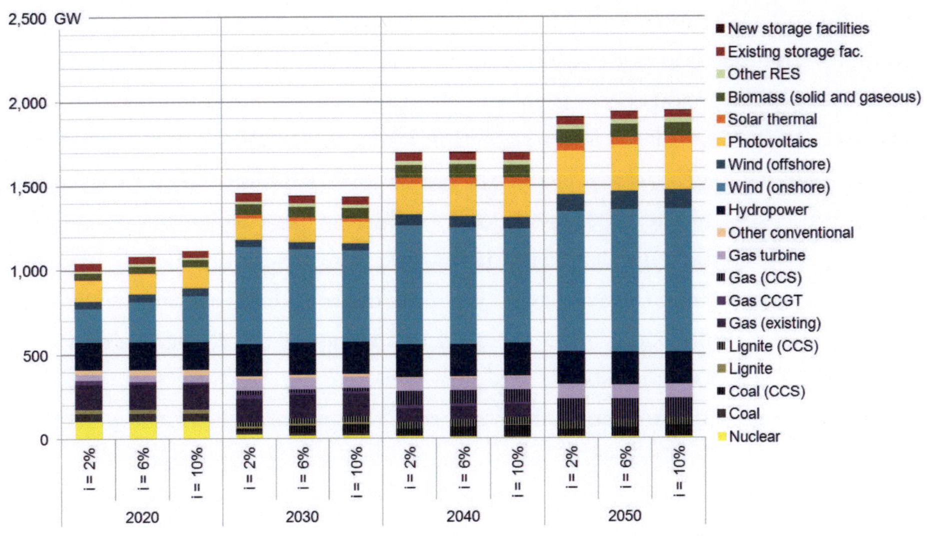

Figure 6.26.: Installed generation capacities for different discount interest rates.

6.3.2. Decarbonisation rate

It is reasonable to assume that the degree to which decarbonisation is imposed on the power sector has a large influence on most results. Therefore, two additional variants of the OPT scenario are calculated, changing the emission cap to levels equal to a 90 % and 98 % reduction compared to 1990 levels. The results discussed in the following focus on the generation mix in 2050 and costs over time, as these exemplify the changes well.

Table 6.4.: Overview of changes for different emission reduction rates in 2050.

	90 %	95 %	98 %	Unit
CO_2 emissions	141.3	75.0	30	Mt
Effective carbon price	129.3	372.9	498.2	EUR/t
Conventional generation	1,422	997	959	TWH
RES-E	3,313	3,738	3,777	TWH
Net RES-E share	70.0	79.0	79.8	%
Average carbon intensity (conventional)	99.4	75.3	31.3	kg/MWh
Average carbon intensity (total)	29.8	15.87	6.38	kg/MWh

Figure 6.27 shows the generation mix over time and central changes for 2050 are summarized in table 6.4. Firstly, has to be pointed out that the "90 % variant" reduces emissions below 90 % of the 1990 level; the implemented carbon prices

alone lead to CO_2 emissions of 141 Mt in 2050, which is equivalent to a reduction of 90.6 %. Lower carbon caps would thus not have any impact without a reduction of the assumed exogenous carbon price.

A general trend is that the lower reduction leads to less RES-E in the generation mix. In 2050, net RES-E share is 70.0 %, 79.0 % and 79.8 % in the three scenarios. This indicates that in the OPT scenario the RES-E is close to its economic limits. For higher decarbonisation rates, the contribution from RE is only slightly increased by the model and instead the average emissions of conventional generation decreases, as can be seen in table 6.4. This is achieved by increasing the capacities of nuclear power and CCGT power plants.[23]

For the less ambitious decarbonisation path, the model reduces the generation from wind power and PV, while solar thermal stays at comparable levels in all variants. The strongest decrease takes place for offshore wind, which is 34 % below the generation in the OPT scenario. This leads to 57 % less curtailed generation potential. In turn, generation from CCS coal power plants increases by 182 %. Although lignite capacities are similar due to the applied restrictions, generation increases by 29 %. The higher utilisation of gas turbines results in 285 % higher generation. CCS gas power plants are the only conventional technology with lower generation, decreasing by 47 %.

With the energy demand and carbon prices of the OPT scenario, the optimal RES-E share seems to be between 70 and 80 % regardless of the carbon cap. This indicates a certain robustness of the result that even with CCS competing in the field of low carbon technologies, the major share of emissions reductions comes from RE in a cost-efficient solution.

The development of the costs is shown in figure 6.28.Compared to the 95 % emission reduction in the four main scenarios, costs decrease by 3.8 % for the less ambitious reduction while increasing by 4.1 % in the more ambitious case. When considering the significant differences in the resulting power systems, these figures appear to be rather low. However, the effective carbon prices in 2050 clearly show the differences in ambition level: while a 90 % reduction can be achieved with a moderate effective carbon price, the value almost triples for a 95 % reduction. For a reduction by 98 %, the value reaches almost 500 EUR/t. This shows that although the power sector offers large reduction potentials, the costs increase sharply when approaching a complete decarbonisation. With the assumed level of electricity de-

---

[23]In reality, capture rates are not necessarily limited to the assumed values, and it might be an alternative approach to move completely to Oxy-fuel CCS power plants, for which capture rate close to 100 % are possible.

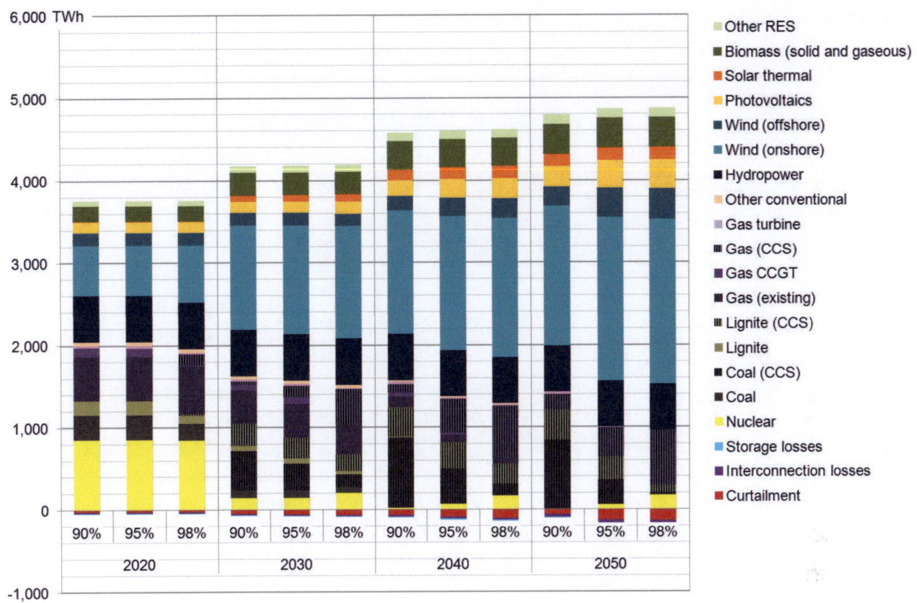

Figure 6.27.: Influence of changes in the imposed emission reduction on the generation mix, with 95 % being the default value in the scenarios.

mand, a reduction by 95 % is already at a relatively steep point of the underlying mitigation cost-potential curve. It could be approximated by exploring mitigation costs systematically for different decarbonisation rates. A combination of such data with its counterparts for the demand side and other sectors could be used to derive the optimal emission reduction rate of the power sector.

6.3.3. Specific investments of gas and nuclear power plants

As explained in the respective sections, the techno-economic assumptions on nuclear and CCS gas power plants are subject to higher uncertainty than other technologies. Therefore, two sensitivity runs are performed with altered specific investments: "Cheap nuclear" and "Expensive gas".

For nuclear power, several sources indicate that the assumed specific investments of 4,000 EUR/MW are in the lower range of plausible parameter space (see chapter 5). However, the conditions in the OPT scenario result in very low new nuclear capacities of 7.4 GW by 2050. Even small increases in specific costs would erase nuclear power from the solution. It seems that the initial parameter selection represents the edge at which nuclear power is part of a cost-efficient generation mix.

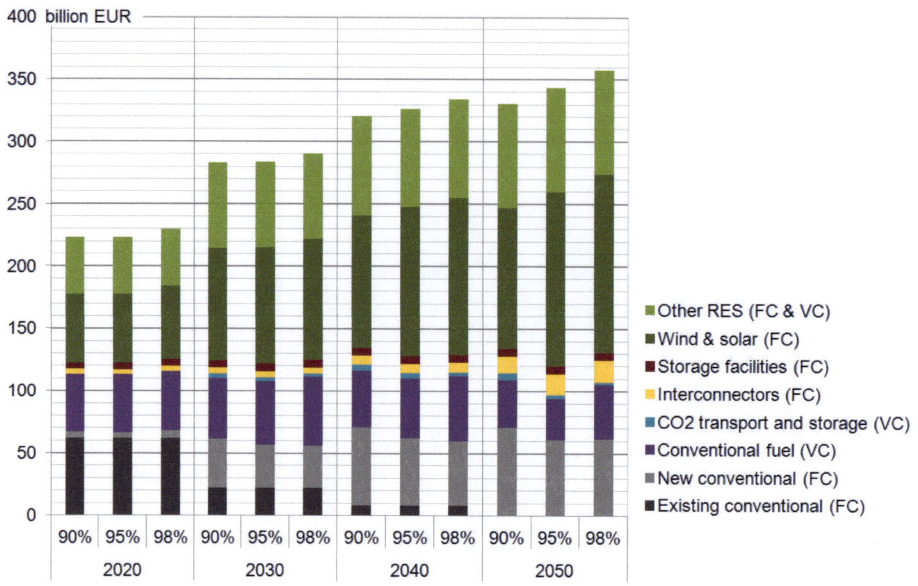

Figure 6.28.: Influence of changes in the imposed emission reduction on the costs, with 95 % being the default value in the scenarios.

However, constructing such a small number of plants seems implausible, because it would be uneconomic to maintain construction facilities and abilities for such a low demand.If nuclear power becomes a part of the decarbonisation portfolio, a substantial role is inevitable. Therefore, the reaction of the model to a lower price of 3,500 EUR/MW has been tested. It has to be noted that this is significantly below the costs that can be expected for the plant currently under construction. However, the applied EPR technology is in an early market stage and such cost reduction could be possible.

In turn, for CCS CCGT power plants the chosen techno-economic assumptions are rather favourable. As the recently published meta-study by Finkenrath (2011, p. 34) suggests, adding carbon capture to CCGT power plants increases costs by approximately 82 % on average. For the sensitivity analyses, an increase of 70 % is assumed in 2020, gradually decreasing to 55 % in 2050.The lower mark-up is chosen as the same study also shows that the price increase is below average in Europe.

The impact of the changes on the generation can be seen in figure 6.29. As it can be seen, the results of the two deviations are very different. In "Cheap nuclear", 59.6 GW of nuclear power is built by 2050, generating 456 TWh. This represents a share in total net generation of 9.6 %. Notably, annual generation is

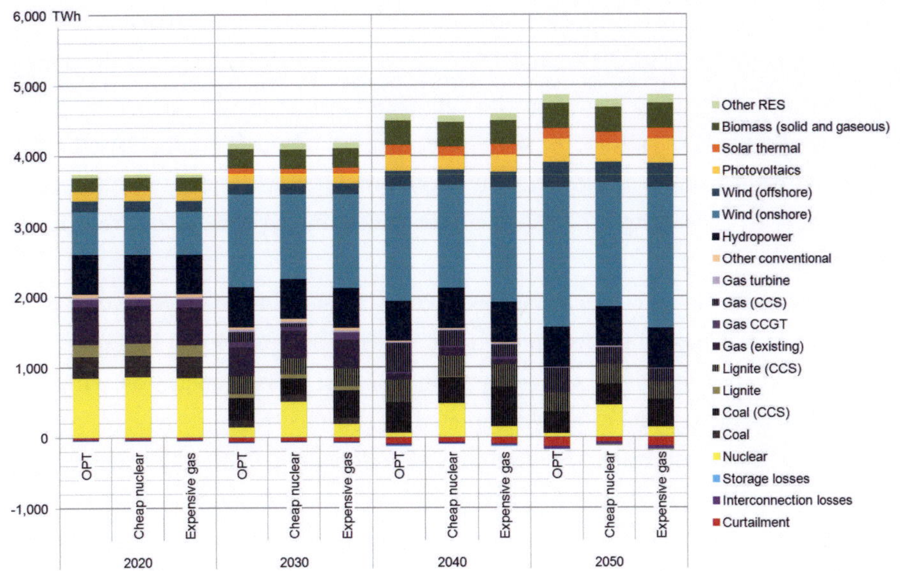

Figure 6.29.: Generation mix with altered specific costs for nuclear energy and CCS CCGT power plants.

at a similar levels from 2030 to 2050. The additional generation predominantly displaces generation from CCS gas power plants[24] (-32 %), PV (-20 %), offshore wind (-17 %) and onshore wind (-12 %). In total, RES-E decreases by 357 TWh. The model especially reduces generation from sites that cause high curtailment in the OPT scenario; this shift decreases curtailment by 47 %.

The assumptions in the "Cheap nuclear" sensitivity run are very optimistic: not only are specific investments very low, the plants also have unlimited flexibility in terms of ramp rates or minimum load, which is impossible in reality. Furthermore, the plants profit from the overnight costs assumption, neglecting the capital cost during the construction phase; in reality the long construction time of nuclear reactors causes high capital costs before the first revenues can be made. Even under these optimistic assumptions, the role of nuclear energy in a decarbonisation scenario remains limited. The result that from an economic point of view nuclear power should not play a significant role with CCS technology as competitor, appears to be robust to changes in the assumptions. However, regional differences in CCS availability, for example if certain regions do not have geological storage potential, could result in regional differences.[25]

---

[24]In turn, generation from open-cycle gas turbines increases by 80 %.

[25]Additional scenarios showed that, the model seeks to sustain countries without CCS potential

The changes in the specific investments of CCS gas power plants have little to no impact on generation from RE; net RES-E share increases by only 0.3 percentage points. Instead, it leads to a shift in the conventional power mix. Generation from CCS gas power plant decreases by 49 %. The missing power is compensated for by higher generation from coal plants. The model maintains the carbon intensity of the power mix by "blending in" nuclear power and reducing generation from lignite power plants.

In summary, the model is much less sensitive to changes in the specific costs of gas power plants than those of nuclear power. Changes in fuel prices can be expected to result in similar changes within the conventional part of power generation. However, as the effect of deviating fuel price is dampened by the effective carbon price, changes would have to be substantial to cause changes equivalent to the "Expensive gas" case. For nuclear power, altered fuel prices, at least within plausible ranges, would have little impact, as fuel costs play a relatively small role in the technologies LCOE.

6.4. Comparison of the results with other studies

Comparing the results of the scenarios to similar research is difficult. Although several studies deal with the decarbonisation of the power sector, the applied models and input parameters are different from the ones used here. The model choice can have an especially significant impact. Recently, a meta study compared several studies on the topic Fischedick et al. (2012). The study covers the following publications:

- The 3rd edition of "energy [revolution. A Sustainable World Energy Outlook" by Greenpeace and EREC (see: Teske, Arthouros Zervos, et al. (2010)).

- "Power Choices - Pathways to Carbon-Neutral Electricity in Europe by 2050" by Eurelectric (2009)

- "Roadmap 2050 - Practical Guide to a Prosperous, Low-Carbon Europe" by the European Climate Foundation (see: ECF (2010))

- "Transformation of Europe's power system until 2050" by McKinsey & Company (2010)

---

through its neighbours; e.g. if Portugal does not have CCS potential, capacities are increased in Spain and the generated power is exported. Over long transportation distances this solution looses its economic attractivity.

All studies are performed with different models. In many aspects, the model employed by ECF has similarities to PowerACE-Europe; however, its capacity expansion is exogenous, making the results difficult to compare. The objective of the study of McKinsey is very similar to the rationale of this thesis, which is to determine the cost-effective way of a 95 % reduction in greenhouse gas emissions from the European power sector. The study employs three different models with a relatively high spatial resolution. The Eurelectric study is modelled with the PRIMES energy model, which is also the model behind the Energy Roadmap of the European Commission.

Because the studies deviate in a large number of details, comparisons beyond the technological "big picture" are probably not worthwhile. Almost all of the covered scenarios have a similar electricity demand as the one assumed in the OPT, NoCCS and GRID scenario. Figure 6.30 shows the supply mix in selected scenarios of the studies, which deviate strongly from the ones presented in this thesis. None of the scenarios results in a similar power mix in 2050 to any of the scenarios discussed.

Firstly, it makes sense to distinguish Greenpeace's (GP) scenarios from the others, as GP does not utilise nuclear energy or CCS. All non-GP studies result in significant proportions of nuclear energy of 12 to almost 50 %. In contrast to this, PowerACE-Europe concludes that nuclear energy is only a part of the cost-efficient solution if CCS does not become available. This is not the case in the non-GP studies. In turn, CCS gas power plant play a role only in the ECF scenarios, in which this diffusion is defined exogenously, and in Eurelectric's "Power Choices" scenario. This is interesting insofar as the latter study also assumes a doubling of pumped storage facilities, which does not seem mandatory in the face of the low proportion of fluctuating RES-E. The scenario has a very high proportion of dispatchable (large-scale) generation capacities.

Figure 6.31 shows RES-E share in the scenarios analysed in the meta-study, as well as the developments in the OPT scenario. Only four scenarios are comparable to the RES-E of this thesis.[26] The highest values are reached for the two GP scenarios, which is self-evident considering the technological options; without the availability of CCS and nuclear energy, RE is the only option for significant decarbonisation. The developments in the "80 % RES" scenario of ECF are, in terms of RES-E share, extremely similar to those in the OPT scenario. However, the RES-share in 2050 of

---

[26]The observable gap between "high RES" and "low RES" scenarios that is softened only by ECF's exogenously set 60 %-RES scenario is remarkable. The gap also seems to exist in many other studies not discussed here; renewables either play the lead role or only a small one. Whether this is a technical result or rather reflects the conviction of the modellers and authors remains to be analysed.

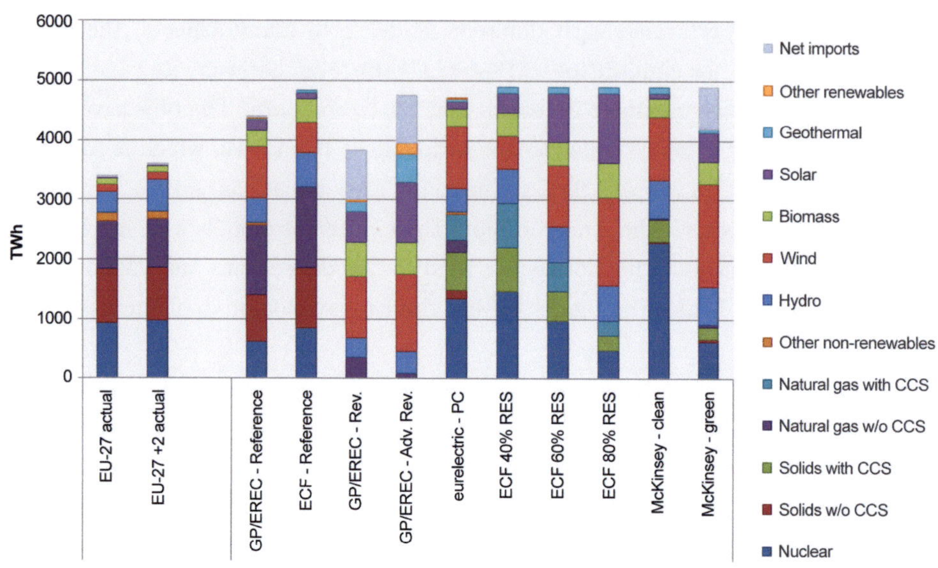

Figure 6.30.: Electricity generation by source (including net imports) in 2008 (actual) and in 2050 according to the different scenarios. Source: Fischedick et al. (2012).

the ECF scenario is not surprising, as it is exogenously set. It is the same situation for McKinsey's "Green" scenario, where a large part of RES-E is imported from solar power in North Africa. Such imports are not included in the scenarios of this thesis; a discussion of a transcontinental integration of Europe, North Africa and the Middle East can be found in Zickfeld and Wieland (2012). The scenarios in the latter publication are also calculated with the model PowerACE-Europe.

In conclusion it can be said that all other studies reaching a similar proportion of RE in the power sector are somewhat predefined. PowerACE-Europe concludes that a high proportion of RES-E is beneficial even if nuclear and energy and CCS are available. For an optimisation model, this is a surprising result. The reason for the differences is not entirely clear, as information on the models behind the other studies is limited.

However, a possible and plausible answer is that PowerACE-Europe depicts the fluctuation of RE better than other models. The high temporal resolution and coverage reveals many issues of RE that affect the technologies themselves, but also the profitability of other technologies. Modelling RES-E on the basis of actual weather data instead of type days often decreases the achievable utilisation of conventional power plants. Arguably, this depends on the accuracy of the applied type day profiles. Depending on how well the actual standard deviation is met by the profile,

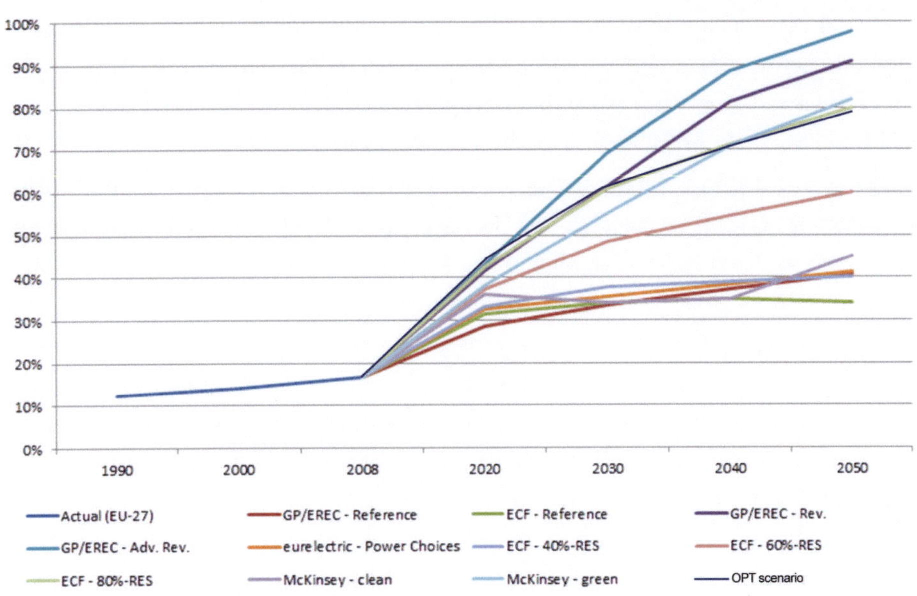

Figure 6.31.: Development of the share of renewable energy sources in electricity generation (including net imports) in the different scenarios. Source: Fischedick et al. (2012), expanded by the OPT scenario.

the opposite could theoretically be true. However, type day profiles underestimate the ability of fluctuating generation to contribute to solving its own issues. When modelling only a small region, the problems of meeting demand with fluctuating RES increase with the level of detail. For a lager regions, the likelihood increases that a too high or too low generation can be compensated for by other regions with the opposite problem. The model uses these option and designs a system in which all components are perfectly matched. [27] Therefore, the higher temporal resolution and coverage of PowerACE-Europe might be a central driver behind the higher RES-E shares that it determines to be optimal.

6.5. Summary and conclusion

This chapter presented the results of four decarbonisation scenarios for the European power sector calculated with the model PowerACE-Europe. The analyses lead

---

[27] Additionally, although this could not be verified for all of the other models compared by the meta-study, some model do not allow curtailment, which would decrease the optimal RES-E shares substantially.

to several conclusions that appear to be robust and applicable for a wide range of possible circumstances. These will be summarised in the following paragraphs.

The most striking result is the high proportion of RES-E computed by the model for all analysed decarbonisation pathways; it constitutes between 79.0 and 81.4 % of total generation in the scenarios. If an almost complete decarbonisation of the power sector is the objective, RE will have to become the central pillar of the system and the generation capacities should be distributed homogeneously over Europe. Between 44 and 54 % of the generated power comes from fluctuating RES. This result is remarkable, especially considering that it is derived with an optimisation model with a very high level of detail in the representation of RE characteristics. The temporal resolution and coverage of the model on the basis of real weather data reveals that many of the technical issues arising from fluctuating supply can be balanced out between regions.

For integrating the high generation from RE, an extensive expansion of the interconnections between the countries is cost-efficient. However, the GRID scenario shows that with CCS available, the most substantial grid expansions can be avoided at a moderate increase of total system costs. Nevertheless, even this scenario implicitly requires a high level of coordination in the planning of all system components. This also means that the decarbonisation of the power supply requires a high willingness to cooperate and exchange power between the European countries. The substantial grid expansion leads to a paradigm shift, in which the *regional residual load* becomes the determining factor for load segment considerations.

The modelling approach also shows that large-scale electricity storage facilities, which are often discussed as a measure for managing the issues of fluctuating supply, do not stand up to closer scrutiny. Even with favourable assumptions, building new storage facilities is only cost-efficient in a few countries and with rather limited capacities. This is the case for both short-term and long-term storage. The reasons for this lie in their direct costs as well as the indirect costs arising from losses, rendering them uncompetitive to other options. This does not change even if there is a high proportion of curtailed generation; in some countries over 20 % of the potential generation from wind and solar power has to be curtailed. This appears to be a characteristic of systems with a large proportion of fluctuating RES but might allow the excess energy to be used in other sectors, e.g. for electric vehicles or hydrogen production for fuel cell vehicles.

If CCS technologies are available, they become an important pillar of the decarbonisation strategy. The proportion of CCS in power generation reaches up to 22.4 % in 2050. Coal and lignite CCS power plants are economically attractive, but their optimal market share depends on their specific emissions, which is subject to the assumed capture rates of the carbon sequestration process. The higher the remaining emissions and the more ambitious the decarbonisation target, the more attractive CCS gas power plants become.

The results indicate that not having CCS available as a decarbonisation option increases the costs covered by the model by 3.8 %[28], which seems substantial, though not exceptional. It increases the necessity for grid expansion and drives up the countries' import dependencies. The main challenge for a power sector without CCS is the lack of dispatchable low-emission generation capacities for the midload segment. However, if CCS is also unavailable in the other sectors or countries as well, the feasibility of the 2°target, that is challenging even with CCS available, can be assumed to decrease strongly.

The results also show that nuclear energy is competitive with CCS only under optimistic techno-economic assumptions. If CCS is not available, nuclear power has to play a significant role, at least in scenarios with high energy demand. In this case, the new nuclear reactors reach very high utilisation despite the high proportion of fluctuating RES-E. This utilisation is enabled by a substantial expansion of interconnection capacities, which are built with consistent strength due to the homogeneous distribution of RE sites.

A lower electricity demand, for example through energy efficiency measures, has a minor influence on the optimal RES-E proportion and generation mix; only the contribution of CCS gas power plants and PV significantly decreases. However, the lower electricity demand decreases the effective carbon price by 65 %, indicating the lower incentives necessary to reach the targeted emission reduction. It also reduces the need for grid expansion; in the light of the issues that the construction of new power lines currently face, this might increase the feasibility of the decarbonisation strategy.

Although the costs of electricity supply differ between the four scenarios analysed, it seems that the only way of substantially decreasing them is to decrease electricity demand. This benefit has to be offset against the demand-side costs of efficiency measures. Nevertheless, even a lower demand does not prevent a rise in specific electricity costs, which increase similarly in all decarbonisation scenarios by up to 26 % compared to 2020. However, in the face of the extensive transformation process towards a decarbonised power supply as described by the scenarios, the financial burden can be seen as moderate.

---

[28]By coincide the same difference as between the 90 % and 95 % emission reduction cases.

7. Summary, conclusions and outlook

7.1. Motivation and objective

Over the last two decades, Europe's electricity sector has changed significantly. This is to a large degree due to the politically forced market liberalisation. Although this process is still ongoing, fighting climate change has emerged as another and possibly more demanding driver of change. To quantify its contribution to reaching the 2 °C target, the EU has set the target to reduce its greenhouse gas (GHG) emissions by 80 % until 2050 compared to 1990 levels (European Commission, 2011a). This goal, if pursued forcefully, requires a fundamental transformation of the electricity sector and a decrease of its emissions close to zero. Three types of technologies are seen as the central options for generating power at low or no emissions: renewable energies (RE), nuclear power and carbon capture and storage (CCS) technologies. Solid academic consensus exists that electricity generated from renewable energy source (RES-E) will play the major role in any cost-efficient decarbonised European power supply system. In this case, a large proportion of electricity will come from fluctuating sources. This would represent an elemental change from today's system, in which most of the power plants are dispatchable.

Most computational electricity system and market models were developed for dealing with the complexity of market processes. In models analysing long-term infrastructure capacity expansions for a large region, the characteristics of RE technologies are represented in a simplified manner. Usually, the typical load behaviour for a limited number of situations is generalised, especially when a large region is considered (i.e. type day approach). On the one hand, this marginalises or neglects certain issues of fluctuating RES, such as long calms; on the other hand, the assumed repetitiveness of situations reduces the options for balancing electricity between weather regions. It has to be assumed that these simplifications limit the realism and accuracy of the models and potentially of the conclusions drawn.

This thesis seeks to analyse concrete options for decarbonising the European power sector, focussing on the supply side and the necessary infrastructure, i.e. grids and electricity storage facilities. The central objective is to improve the understanding of the role of RE in decarbonising the sector.

7.2. Approach and procedure

In order to pursue the objective of this thesis, an analysis of four different scenarios for decarbonising the European power sector is performed. A model is developed to generate the scenarios, allowing the optimisation of long-term capacity expansion and utilisation and specifically addressing the characteristics and impacts of fluctuating RES-E.

The work starts with a summary of the recent developments in the European power sector. The goal is to provide an overview of the implications that current regulation and trends will have for future decades and for approaches to model the sector. It is concluded that despite several drawbacks, especially in the early years, EU energy policy guides the system towards an internal market for electricity and counteracts market barriers such as lack of transparency. If this process continues the error caused by assuming a perfect market and perfect competition, which are presumed in most power system models, is likely to decrease over subsequent decades. Several political measures directly or indirectly influencing the decarbonisation of the power sector can be seen as exogenous influence and they have to be adequately considered. Among these regulations, the EU Emissions Trading System (ETS), the binding RE targets for 2020 and several national nuclear power phase-out policies currently seem to be the most influential.

The next step introduces the main modelling approaches applied in analysing the electricity sector. *Electricity market models* focus on the effects of imperfect markets as well as players' strategies and interdependencies. The level of technical detail varies significantly, from top-down macroeconomic models to bottom-up approaches such as agent-based simulation. While these models often deliver realistic results, especially in short- to mid-term horizons, they come with certain limitations in terms of size of analysed region, time horizon or resolution. By contrast, *electricity system models*, such as optimisation models, typically feature a less detailed depiction of market rules and processes, but allow for the analysis of large systems over longer time horizons. The comparison shows that no model concept is superior in all aspects but all have particular strength and weaknesses.

Therefore, the model capabilities necessary for the task at hand are defined and existing models are assessed on this basis. In the face of a growing proportion of RES-E, an hourly temporal resolution seems necessary, and a large number of hours need to be covered over the years; when analysing future systems based to a large extent on wind and solar power, simplifying their complex characteristics to a few "typical" states neglects some challenges and possibilities alike. A comparison of existing and available models leads to the conclusion that there is no long-term capacity expansion model for large regions which provides sufficient time coverage. In the cases where the existing models are theoretically capable of hourly resolution throughout entire years, the consequences in terms of model run-time are uncertain. Therefore, a new model is developed.

Instead of creating the model from scratch, the infrastructure of the existing electricity market model cluster *PowerACE* is used, one version of which is maintained by Fraunhofer ISI. The cluster's main model is an agent-based unit commitment model focusing on Germany. The model's efficient data management is used, but the actual core model parts, i.e. the generation capacity expansion planning and the unit commitment components, are re-developed from scratch.

The new electricity system model *PowerACE-Europe* follows the optimisation approach; it seeks a cost-efficient solution for meeting electricity demand under several restrictions, mostly of a technical nature. The model's central strength is an hourly resolution in long-term scenarios, optimising the years 2020, 2030, 2040 and 2050 simultaneously. For this thesis, it covers the EU-27 along with Norway and Switzerland. The optimisation comprises both capacity expansion and utilisation for conventional power generation, wind and solar technologies, interconnecting electricity grid between countries as well as electricity storage facilities. Fluctuating RES-E is modelled on the basis of actual weather data, thus incorporating correlations between wind and solar power as well as between weather regions. The optimal solution is found by solving a large linear problem through interior-point optimisation, which is performed with the CPLEX solver. The scenarios can be set up to meet additional requirements, such as emission reduction targets.

7.3. Scenario definition and main results

The scenarios examined in this study are designed alongside uncertainties that are assumed to have an immense impact on future developments in the power sector; they address the consequences of different developments in the context of these three currently urgent questions:

1. What are the consequences if CCS technologies do not become available?

2. Will the development of the grid be fast enough for the transition necessary for the decarbonisation of the power sector?

3. Will efficiency measures be able to significantly decrease electricity demand?

The scenarios to some degree represent a trade-off between consistent and coherent developments, e.g. taking into account the interdependencies between fuel consumption and prices, whilst keeping the differences between the scenarios at levels at which changes in results can be traced back to their causes. While in all scenarios CO_2 emissions from the power sector are reduced by 95 % compared to 1990 levels, table 7.1 shows the differences in input assumptions.

Table 7.1.: Overview of the main differences between the analysed scenarios .

	Optimistic Decarbonisation	No CCS	Hampered Grid	Strengthened Efficiency
Abbreviation	OPT	NoCCS	GRID	EFF
Electricity demand	high	high	high	low
CCS technology available	yes	no	yes	yes
Prompt grid extensions	yes	yes	no	yes

For the four scenarios the cost-efficient solution is determined with PowerACE-Europe. The analysis of the results leads to several conclusions that are assessed to be robust and applicable for many developments not covered by the scenarios.

The most striking result from the model is the conclusion that a high proportion of RES-E is optimal in all analysed cases, lying in the range of 79.0 to 81.4 %. If an almost complete decarbonisation of the power sector is the objective, RE has to become the central pillar of the system. The model distributes the generation capacities of both RE and conventional plants relatively homogeneously over Europe. Between 44 and 54 % of the generated power comes from fluctuating RES. This result is remarkable insofar as it is determined by an economic optimisation model with a very high level of detail in the representation of RE characteristics. The temporal resolution and coverage of the model on the basis of real weather data suggests that many of the technical issues arising from fluctuating supply can be balanced out between regions.

The infrastructure components have to be well matched and an extensive expansion of the interconnections between the countries is necessary. The strength[1] of the interconnectors increases by 182 to 507 % by 2050 depending on the scenario; the highest value occurring if CCS technologies are not available. In the GRID

---

[1] Grid strength is measured in this thesis as the cumulative product of lengths and net transfer capacity of all interconnectors, expressed in GW km.

scenario, in which the grid expansion is exogenously set to be 10 years behind the optimal development, costs in 2050 increase only moderately by 2 billion EUR[2]. In this scenario, the model increases the flexibility by expanding CCS gas power plant capacities. However, it has to be mentioned that the model, like all optimisation models, is not able to reveal all issues arising from grid expansion delays. Its perfect foresight allows it to anticipate the bottlenecks and balance the system cost-efficiently using the remaining degree of freedom.

New large-scale electricity storage facilities, which are often discussed as a measure for managing the issues of fluctuating supply, are cost-efficient only under special cases; installed peak capacities increase by 4.2 GW (EFF scenario) to 16.9 GW (NoCCS) scenario; this equals an increase by a maximum of 38 %. The storage only reaches utilisation that justifies its high fixed costs only in a few locations. This result seems counter-intuitive, since up to 5 % of the potential generation from wind and photovoltaic power plants is not utilised and has to be curtailed. This appears to be a general characteristic of systems with high a proportion of fluctuating generation. This might offer potential to use the excess energy in other sectors, e.g. for electric vehicles or for hydrogen production for fuel cell vehicles.

If CCS technologies are available, which in reality also requires the solution of non-technical issues, such as social acceptance problems, they become an important pillar of the decarbonisation strategy. The technology's share in power generation in 2050 reaches up to 22.4 % in the scenarios. Due to the explicit or implicit costs of emissions, coal and lignite CCS power plants are economically attractive in the long run. Their optimal market share depends to a significant extent on the specific emissions, which are subject to the capture rates of the carbon sequestration process. Not having CCS available, *ceteris paribus*, substantially increases the costs in 2050 by 13.1 billion EUR, which equals 3.8 % of the costs covered by the model. It drives up the necessity for grid expansion and causes countries with low RE potentials and a nuclear phase-out policy to be relatively import-dependent. The central challenge for a power sector without CCS is the lack of dispatchable low-carbon generation capacities for the mid-load segment.

The results also show that nuclear energy is only competitive with CCS under optimistic assumptions. Nuclear energy only plays a significant role if CCS is not available and energy demand is high. Under these conditions, new nuclear reactors with a total capacity of 112 GW are built up until 2050; the plants have a high utilisation despite the significant proportion of fluctuating RES-E. This is possible

---

[2]Monetary figures are stated as real values in EUR$_{2010}$.

through the immense expansion of interconnection capacities, which often allows export when RES-E exceeds domestic demand.

A lower electricity demand, e.g. through energy efficiency measures, has only a minor influence on the optimal RES-E share and generation mix; only the contribution of CCS gas power plants and PV significantly decrease. Nevertheless, the lower electricity demand decreases the shadow price of the CO_2 emission constraint, indicating the lower "incentives" necessary to reach the enforced emission reduction. A lower demand also substantially reduces the need for grid expansion; in the light of the challenges that the construction of new power lines currently face, this might increase the feasibility of the decarbonisation strategy.

Although the absolute costs of electricity supply differ between the four analysed scenarios, specific costs are relatively comparable. There is an increase from approximately 60.2 EUR/MWh in 2020 to 72.2 in the EFF scenario and to 76.1 EUR/MWh in the NoCCS scenario; the latter equals an increase by up to 26 % between 2020 and 2050.

This benefit has to be offset against the demand-side costs of efficiency measures. However, even a lower demand does not prevent a rise in specific electricity costs, which similarly increase in a decarbonisation scenario by up to 26 % compared to 2020. In the face of the extensive transformation process necessary for an almost complete decarbonisation, the financial burden appears to be moderate.

Subsequently, a sensitivity analysis is performed for key input parameters. While the influence of the interest rate applied for discounting is relatively small, the level of emission reduction has naturally far-reaching implications. A comparison reveals that with the high electricity demand assumed in three of the four scenarios, an emission reduction to 95 % of 1990 levels is already at a steep point in the emission abatement cost curve; compared to a reduction by only 90 % the shadow price of the emission constraint in 2050 almost triples.

7.4. Methodological evaluation and outlook

For this thesis, the power system optimisation model PowerACE-Europe has been developed. While the approach in general is not a novelty, the model combines features and capabilities that facilitate a better understanding of the interdependencies between system components. The model's central strength is its high spatial and temporal resolution and coverage in combination with a level of substantial technical detail. Few power system models exist that are able to optimise capacity expansions whilst taking into account all hours of the scenario years; to the knowledge of the author, all models with similar temporal coverage analyse only a single country. By using actual weather data for the generation of RE load profiles, many meteorological phenomena are implicitly included. The combination of these features allows integrated examination of conventional and renewable power generation, interconnections and storage facilities in systems with high proportions of RE. The solutions the model generates for the scenarios in this thesis endorse this approach; they are possible only through a detailed modelling of the fluctuations over a large area, partially balancing each other out.

The drawback is that the approach uses and generates excessive amounts of data, requiring very advanced hardware to handle it; even with very powerful high-end servers, a typical run takes over a day to finish. This issue limits the room for further model improvements in the short-term. However, a logical next step is the inclusion of supply side activities, starting by taking into account future changes in the load profile through new applications or altered consumption behaviour. It seems plausible that integrating demand response options, for example from heat pumps or electric vehicles, will have a significant impact on the results.

The range of further potential research questions to be assessed with the model is large. Currently, its geographical region is being expanded to include Northern Africa and the Middle East. This allows assessing the potential benefit of exchanging power in a wider region or importing power generated in the deserts to Europe. The inclusion of demand response or efficiency measures would make it possible to quantify the benefits to be gained and to design demand-side incentives accordingly.

In order to further expand the possible fields of application for the model, its integration and interaction with the other components of the PowerACE cluster, or other models in general, should be advanced. This would partially compensate for the shortcomings inherent in the modelling approach itself. For example, if the solution of the optimisation model in terms of installed capacities was fed back into the agent-based model of the cluster, the performance of the infrastructure under particular market rules could be assessed. This would open the model to questions of market design, such as the potential necessity of capacity markets. The model represents a step towards developing technically feasible solutions for decarbonising the power sector through high proportions of renewable energies; improving the understanding of concrete ways for implementing such solutions in real-world markets is necessary and will remain a challenging field of research.

A. Additional tables

Table A.1.: Annual generation from RE in TWh, calculated by the agent-based investment diffusion model *ResInvest*. Wind and solar power are included as additional information, though the data is not used in PowerACE-Europe.

Technology	2020	2030	2040	2050
Biogas	67,860	107,492	138,335	144,862
Biomass	134,576	201,566	243,726	255,479
Biowaste	14,047	14,443	14,636	14,678
Geothermal	6,054	6,054	6,207	6,955
Hydropower	565,256	565,469	566,139	566,268
Landfill gas	25,232	27,643	27,685	27,698
Sewage gas	3,818	4,834	5,602	5,682
Photovoltaics	137,419	339,762	494,612	598,320
Solar Thermal	5,853	20,281	29,357	33,769
Tidal	2,209	4,815	7,003	12,317
Wave	8,508	24,937	37,210	48,873
Wind (onshore)	153,091	277,196	453,402	538,915
Wind (offshore)	498,787	1,126,813	1,835,750	2,105,082
Total	1,622,710	2,721,304	3,859,663	4,358,899

Table A.2.: Electricity generation by type and usage in the OPT scenario in GWh.

	2020	2030	2040	2050
Demand, losses and curtailment				
Net demand and interior losses	-3,705,754	-4,097,515	-4,469,344	-4,686,238
Interconnection losses	-14,245	-17,060	-24,862	-41,554
Storage losses	-6,197	-8,075	-12,687	-7,000
Curtailment	-26,151	-56,867	-89,214	-131,127
Generation				
Nuclear	848,786	141,249	66,029	57,806
Coal	301,087	94,240	3,576	0
Coal (CCS)	4,385	324,696	424,497	299,273
Lignite	167,481	59,668	1,497	0
Lignite (CCS)	3	256,505	320,209	277,912
Gas (existing)	539,149	410,866	88,848	0
Gas turbine	24,185	26,415	15,870	7,681
Gas (CCGT)	106,385	72,507	25,343	0
Gas (CCS)	2	144,892	407,497	353,844
Other Conventional	48,145	37,128	16,360	0
Hydropower	564,267	565,467	565,140	566,318
Wind (onshore)	608,882	1,319,031	1,628,494	1,985,903
Wind (offshore)	152,694	152,694	220,725	350,604
Biomass (solid and gaseous)	187,614	271,854	343,771	362,012
Photovoltaics	133,916	134,128	226,713	332,704
Solar thermal	5,756	85,431	143,593	155,631
Other RES	59,703	82,725	98,075	116,206
Total generation (potential)	3,752,438	4,179,496	4,596,236	4,865,893

Table A.3.: Generation capacities installed in the OPT scenario.

	2020	2030	2040	2050
Nuclear	103,436	18,073	8,126	7,396
Coal	44,716	15,878	2,128	0
Coal (CCS)	527	40,609	55,811	56,479
Lignite	23,158	10,533	2,170	0
Lignite (CCS)	0	32,015	41,174	42,160
Gas (existing)	152,883	139,410	78,963	0
Gas turbine	41,402	69,173	77,410	85,502
Gas (CCGT)	14,340	14,340	14,340	0
Gas (CCS)	0	22,696	80,914	126,697
Other Conventional	29,226	14,825	6,051	0
Hydropower	163,771	192,219	192,436	192,466
Wind (onshore)	237,090	551,381	690,575	845,006
Wind (offshore)	45,266	45,266	65,497	108,920
Biomass (solid and gaseous)	41,118	60,530	75,094	78,826
Photovoltaics	123,959	124,097	190,296	272,039
Solar thermal	1,420	23,191	41,832	46,248
Other RES	12,608	19,278	23,618	28,312
Total without storage	1,034,920	1,397,411	1,652,073	1,895,905
Storage facilities	44,358	48,255	49,996	50,212
Total including storage	1,079,278	1,441,769	1,696,431	1,940,263

Table A.4.: Electricity generation by type and usage in the NoCCS scenario in GWh.

	2020	2030	2040	2050
Demand, losses and curtailment				
Net demand and interior losses	-3,705,754	-4,097,515	-4,469,344	-4,686,238
Interconnection losses	-14,121	-17,724	-30,973	-61,460
Storage losses	-6,060	-7,167	-11,640	-17,022
Curtailment	-25,445	-70,231	-87,104	-87,476
Generation				
Nuclear	848,844	475,662	796,612	785,455
Coal	299,990	75,374	991	1
Coal (CCS)	0	0	0	0
Lignite	166,952	44,091	658	0
Lignite (CCS)	0	0	0	0
Gas (existing)	528,414	394,726	144,967	0
Gas turbine	21,610	20,027	13,649	24,561
Gas (CCGT)	129,327	415,179	260,896	185,383
Gas (CCS)	0	0	0	0
Other Conventional	48,028	36,913	16,237	0
Hydropower	564,268	565,467	565,140	566,315
Wind (onshore)	604,359	1,385,350	1,579,454	1,920,624
Wind (offshore)	152,694	174,611	350,148	375,796
Biomass (solid and gaseous)	187,614	271,854	343,770	362,013
Photovoltaics	133,916	153,920	274,154	360,555
Solar thermal	5,756	96,725	154,474	155,272
Other RES	59,703	82,725	98,075	116,206
Total generation (potential)	3,751,474	4,192,624	4,599,225	4,852,182

Table A.5.: Generation capacities installed in the NoCCS scenario.

	2020	2030	2040	2050
Nuclear	103,436	62,248	109,983	112,246
Coal	44,716	15,878	2,128	0
Coal (CCS)	0	0	0	0
Lignite	23,158	10,533	2,170	0
Lignite (CCS)	0	0	0	0
Gas (existing)	152,883	139,410	78,963	0
Gas turbine	38,903	60,751	67,928	114,774
Gas (CCGT)	17,560	62,191	73,031	74,611
Gas (CCS)	0	0	0	0
Other Conventional	29,226	14,825	6,051	0
Hydropower	163,771	192,219	192,436	192,466
Wind (onshore)	235,134	578,722	662,105	803,931
Wind (offshore)	45,266	51,547	109,027	116,974
Biomass (solid and gaseous)	41,118	60,530	75,094	78,826
Photovoltaics	123,959	138,032	227,647	293,276
Solar thermal	1,420	26,618	45,854	46,479
Other RES	12,608	19,278	23,618	28,312
Total without storage	1,033,158	1,432,782	1,676,035	1,861,895
Storage facilities	44,358	51,060	56,701	61,248
Total including storage	1,077,516	1,483,842	1,732,736	1,923,143

Table A.6.: Electricity generation by type and usage in the GRID scenario in GWh.

	2020	2030	2040	2050
Demand, losses and curtailment				
Net demand and interior losses	-3,705,754	-4,097,515	-4,469,344	-4,686,238
Interconnection losses	-13,174	-14,822	-20,419	-25,915
Storage losses	-6,278	-8,402	-13,217	-7,269
Curtailment	-36,763	-59,394	-91,802	-124,999
Generation				
Nuclear	847,615	153,466	88,582	95,847
Coal	298,376	93,320	1,307	0
Coal (CCS)	4,293	303,633	401,429	272,639
Lignite	166,785	58,609	1,313	0
Lignite (CCS)	12	260,190	311,860	269,278
Gas (existing)	541,864	410,290	96,115	0
Gas turbine	25,849	28,000	16,169	8,476
Gas (CCGT)	117,665	77,272	26,843	0
Gas (CCS)	5	176,568	452,533	516,410
Other Conventional	48,018	37,128	16,365	0
Hydropower	564,268	565,467	565,141	566,321
Wind (onshore)	607,630	1,291,203	1,582,457	1,802,119
Wind (offshore)	152,694	152,694	220,724	358,337
Biomass (solid and gaseous)	187,614	271,854	343,771	362,011
Photovoltaics	133,915	134,120	231,714	320,427
Solar thermal	5,756	83,584	140,516	156,335
Other RES	59,703	82,725	98,075	116,206
Total generation (potential)	3,762,061	4,180,122	4,594,914	4,844,398

Table A.7.: Generation capacities installed in the GRID scenario.

	2020	2030	2040	2050
Nuclear	103,436	19,584	10,892	12,180
Coal	44,716	15,878	2,128	0
Coal (CCS)	516	37,949	52,921	53,521
Lignite	23,158	10,533	2,170	0
Lignite (CCS)	1	32,435	40,088	41,075
Gas (existing)	152,883	139,410	78,963	0
Gas turbine	43,724	72,336	80,333	81,597
Gas (CCGT)	15,845	15,845	15,845	0
Gas (CCS)	0	27,715	90,520	162,917
Other Conventional	29,226	14,825	6,051	0
Hydropower	163,771	192,219	192,436	192,466
Wind (onshore)	236,700	540,154	672,719	769,933
Wind (offshore)	45,266	45,266	65,497	111,048
Biomass (solid and gaseous)	41,118	60,530	75,094	78,826
Photovoltaics	123,959	124,092	193,773	263,944
Solar thermal	1,420	22,709	40,716	46,406
Other RES	12,608	19,278	23,618	28,312
Total without storage	1,038,347	1,395,746	1,651,051	1,851,224
Storage facilities	44,358	49,346	51,645	53,357
Total including storage	1,082,705	1,440,104	1,695,409	1,895,582

Table A.8.: Electricity generation by type and usage in the EFF scenario in GWh.

	2020	2030	2040	2050
Demand, losses and curtailment				
Net demand and interior losses	-3,676,754	-3,850,768	-3,774,066	-3,367,016
Interconnection losses	-13,667	-16,191	-22,400	-30,267
Storage losses	-6,207	-7,773	-10,122	-8,138
Curtailment	-26,073	-52,302	-77,948	-66,306
Generation				
Nuclear	850,018	136,741	20,591	3
Coal	299,621	96,120	7,260	0
Coal (CCS)	3,639	346,836	370,720	282,528
Lignite	152,295	59,234	4,040	0
Lignite (CCS)	2	205,891	262,094	264,593
Gas (existing)	509,056	382,911	123,376	0
Gas turbine	24,733	20,719	16,348	16,706
Gas (CCGT)	88,079	65,149	31,576	03
Gas (CCS)	2	59,319	149,895	70,050
Other Conventional	47,999	37,115	16,297	0
Hydropower	564,267	565,467	565,140	566,320
Wind (onshore)	643,415	1,250,123	1,378,078	1,289,713
Wind (offshore)	152,694	152,694	181,946	181,946
Biomass (solid and gaseous)	187,614	271,854	343,771	362,012
Photovoltaics	133,916	133,921	194,309	181,785
Solar thermal	5,744	60,184	121,125	139,844
Other RES	59,703	82,725	98,075	116,206
Total generation (potential)	3,722,799	3,927,004	3,884,639	3,471,709

Table A.9.: Generation capacities installed in the EFF scenario.

	2020	2030	2040	2050
Nuclear	103,436	17,491	2,584	0
Coal	44,716	15,878	2,128	0
Coal (CCS)	438	44,151	49,402	49,402
Lignite	23,158	10,533	2,170	0
Lignite (CCS)	0	25,901	33,516	33,516
Gas (existing)	152,883	139,410	78,963	0
Gas turbine	44,151	61,242	67,249	70,524
Gas (CCGT)	12,316	12,316	12,316	0
Gas (CCS)	0	9,725	30,639	32,943
Other Conventional	29,226	14,825	6,051	0
Hydropower	163,771	192,219	192,436	192,466
Wind (onshore)	248,172	514,892	569,172	527,588
Wind (offshore)	45,266	45,266	53,685	53,685
Biomass (solid and gaseous)	41,118	60,530	75,094	78,826
Photovoltaics	123,959	123,963	166,504	158,239
Solar thermal	1,420	16,328	34,246	40,546
Other RES	12,608	19,278	23,618	28,312
Total without storage	1,046,638	1,328,123	1,403,948	1,270,237
Storage facilities	44,358	48,533	48,533	48,548
Total including storage	1,090,996	1,372,481	1,448,306	1,314,595

Table A.10.: Annual electricity demand in the OPT, NoCCS and GRID scenario in TWh/a; values are stated in net electricity demand including internal grid losses of 6.5 %.

Country	2020	2030	2040	2050
AT	61,273	64,409	69,849	73,762
BE	81,710	90,552	104,989	117,065
BG	34,491	41,960	51,293	57,927
CH	84,327	90,058	96,482	101,050
CY	8,052	9,913	10,718	10,777
CZ	77,800	89,826	104,112	111,518
DE	574,336	612,282	648,596	670,524
DK	42,697	47,797	53,833	58,340
EE	9,573	11,563	13,676	14,638
ES	316,736	338,400	367,499	379,960
FI	89,929	94,041	99,762	104,992
FR	532,740	580,696	598,766	613,605
GR	70,734	74,317	77,934	82,948
HU	49,469	58,303	64,767	66,373
IE	31,997	35,635	40,064	46,426
IT	374,962	416,423	452,950	474,357
LT	12,051	14,556	17,215	18,426
LU	6,461	7,161	8,302	9,257
LV	8,786	10,612	12,551	13,434
MT	2,893	3,562	3,851	3,872
NL	134,153	147,584	160,265	168,561
NO	116,429	118,662	122,901	127,687
PL	179,760	230,204	273,040	284,767
PT	70,484	84,549	93,427	98,219
RO	87,119	113,411	151,926	169,255
SE	137,097	144,164	152,239	159,052
SI	16,649	19,374	21,994	22,791
SK	58,674	72,311	79,184	77,338
UK	444,397	486,271	529,225	561,967
Total	3,715	4,108	4,481	4,698

Table A.11.: Annual electricity demand in the EFF scenario in TWh/a; values are stated in net electricity demand plus "internal grid losses" of 6.5 %. Own calculations based on DLR (2006).

Country	2020	2030	2040	2050
AT	66,429	65,711	59,885	48,987
BE	93,344	91,321	82,685	67,027
BG	28,105	28,718	29,177	26,502
CH	64,323	60,819	52,193	39,370
CY	4,743	5,364	5,502	4,936
CZ	60,182	61,265	59,335	51,721
DE	640,324	654,415	625,867	548,817
DK	48,798	52,415	53,236	51,118
EE	7,546	8,784	9,659	9,400
ES	299,064	334,559	345,774	320,107
FI	83,941	84,745	82,355	76,396
FR	542,370	550,231	513,127	425,990
GR	62,425	67,981	68,673	62,140
HU	40,457	44,852	46,872	43,871
IE	35,060	38,365	38,152	33,967
IT	372,815	383,166	362,645	310,646
LT	9,733	11,331	12,460	12,125
LU	10,314	11,226	11,414	10,935
LV	7,132	8,303	9,130	8,885
MT	2,943	3,076	2,877	2,327
NL	131,432	136,942	132,150	116,049
NO	132,647	130,912	123,883	112,049
PL	153,245	178,401	196,172	190,905
PT	53,648	60,510	64,069	62,016
RO	57,786	74,629	91,214	96,132
SE	160,711	163,790	161,739	153,665
SI	12,095	12,092	11,275	9,349
SK	28,227	30,491	31,567	29,458
UK	476,870	506,772	501,185	451,228
Total	3,686	3,861	3,784	3,376

Table A.12.: Cost developments in the scenarios. Value stated in million EUR$_{2010}$ as variable costs (VC) and annualised fixed costs (FC).

	2020	2030	2040	2050
OPT				
Existing conventional (FC)	61,882	21,984	7,907	0
New conventional (FC)	4,597	34,482	54,160	60,462
Conventional fuel (VC)	46,566	51,369	48,251	33,295
CO2 transport and storage (VC)	17	2,770	3,921	3,145
Interconnectors (FC)	4,143	5,206	7,772	16,873
Storage facilities (FC)	5,381	5,818	6,014	6,038
Wind & solar (FC)	54,721	93,856	119,401	140,344
Other RES (FC & VC)	45,892	68,424	79,243	83,399
Sum	223,199	283,909	326,670	343,557
NoCCS				
Existing conventional (FC)	61,882	21,984	7,907	0
New conventional (FC)	4,662	36,573	73,638	80,693
Conventional fuel (VC)	46,711	50,919	33,048	20,779
CO2 transport and storage (VC)	0	0	0	0
Interconnectors (FC)	4,156	5,704	9,800	23,362
Storage facilities (FC)	5,381	6,133	6,766	7,277
Wind & solar (FC)	54,497	100,470	130,234	141,176
Other RES (FC & VC)	45,892	68,424	79,243	83,399
Sum	223,181	290,207	340,636	356,686
GRID				
Existing conventional (FC)	61,882	21,984	7,907	0
New conventional (FC)	4,928	35,816	56,242	67,709
Conventional fuel (VC)	47,156	53,061	51,093	42,883
CO2 transport and storage (VC)	17	2,758	3,864	3,273
Interconnectors (FC)	3,861	4,316	6,007	9,216
Storage facilities (FC)	5,381	5,941	6,199	6,391
Wind & solar (FC)	54,676	92,534	117,569	132,717
Other RES (FC & VC)	45,892	68,424	79,243	83,399
Sum	223,792	284,835	328,124	345,589
EFF				
Existing conventional (FC)	61,882	21,984	7,907	0
New conventional (FC)	4,499	30,276	38,009	37,206
Conventional fuel (VC)	44,319	44,251	32,567	15,604
CO2 transport and storage (VC)	14	2,467	2,992	2,526
Interconnectors (FC)	4,148	5,099	7,031	10,908
Storage facilities (FC)	5,381	5,849	5,850	5,851
Wind & solar (FC)	55,988	88,228	100,561	87,725
Other RES (FC & VC)	45,892	68,424	79,243	83,399
	222,123	266,579	274,160	243,219

Table A.13.: Import dependencies and net import of the countries in the NoCCS scenario

Country	Import dependency				Net import [GWh]	
	2020	2030	2040	2050		2050
LV	-100%	-66%	-68%	-220%	-	29,424
SI	21%	9%	-98%	-199%	-	45,323
EE	21%	4%	1%	-144%	-	20,990
NO	-28%	-46%	-64%	-112%	-	142,019
IE	-11%	-2%	-15%	-81%	-	37,498
DK	-69%	-40%	-22%	-41%	-	23,695
AT	-27%	-38%	-38%	-36%	-	26,732
CZ	17%	-5%	-44%	-35%	-	38,765
GR	-1%	0%	-5%	-32%	-	26,151
FR	-15%	-7%	-16%	-21%	-	129,546
HU	19%	4%	-1%	-18%	-	12,087
SE	-3%	-4%	-13%	-8%	-	12,144
PL	3%	1%	2%	-7%	-	20,481
RO	0%	0%	-1%	-5%	-	9,132
ES	-1%	-1%	0%	-1%	-	5,435
PT	7%	4%	3%	-1%	-	632
CY	0%	0%	0%	0%		0
MT	0%	0%	0%	0%		0
BG	-12%	-7%	-5%	1%		702
UK	2%	-1%	1%	5%		25,494
LT	-15%	28%	11%	7%		1,242
SK	1%	-2%	0%	7%		5,410
NL	6%	7%	6%	12%		20,346
FI	18%	16%	6%	17%		17,424
IT	3%	-1%	9%	28%		134,495
DE	6%	11%	21%	29%		195,050
CH	38%	41%	45%	49%		49,058
BE	21%	10%	35%	54%		62,518
LU	53%	33%	40%	74%		6,855

Table A.14.: Import dependencies and net import of the countries in the GRID scenario

Country	Import dependency				Net import [GWh]
	2020	2030	2040	2050	2050
NO	-27%	-35%	-48%	-60%	- 76,034
EE	8%	-3%	3%	-38%	- 5,596
AT	-27%	-27%	-22%	-29%	- 21,250
LV	-46%	-60%	-34%	-28%	- 3,724
IE	0%	0%	0%	-20%	- 9,046
DK	-50%	-40%	-25%	-19%	- 10,936
LT	-14%	34%	14%	-9%	- 1,726
GR	-1%	-2%	-2%	-9%	- 7,370
BG	-11%	-9%	-6%	-8%	- 4,710
CZ	16%	-9%	-8%	-7%	- 7,604
SE	-3%	0%	-1%	-3%	- 4,574
ES	-1%	0%	3%	-2%	- 8,045
PT	7%	3%	-3%	-1%	- 924
IT	2%	2%	3%	-1%	- 3,627
CY	0%	0%	0%	0%	0
MT	0%	0%	0%	0%	0
RO	1%	1%	0%	0%	532
FR	-15%	-5%	-3%	1%	3,366
PL	2%	1%	3%	2%	4,413
UK	2%	0%	0%	2%	10,398
NL	6%	4%	12%	4%	6,212
DE	5%	5%	0%	4%	29,685
HU	14%	7%	4%	8%	5,201
BE	21%	4%	0%	10%	12,116
SK	-1%	5%	6%	11%	8,215
SI	21%	14%	21%	13%	2,995
FI	18%	16%	12%	18%	19,342
LU	66%	34%	42%	21%	1,985
CH	38%	31%	31%	35%	34,790

Table A.15.: Import dependencies and net import of the countries in the EFF scenario

Country	Import dependency				Net import [GWh]
	2020	2030	2040	2050	2050
LV	-129%	-109%	-89%	-107%	- 9,453
NO	-25%	-34%	-54%	-71%	- 78,740
AT	-17%	-22%	-31%	-55%	- 26,988
IE	-5%	-2%	-3%	-33%	- 11,345
DK	-54%	-33%	-24%	-33%	- 17,023
CH	19%	17%	6%	-23%	- 9,142
BG	-22%	-17%	-11%	-18%	- 4,727
LT	-41%	33%	0%	-17%	- 2,085
GR	2%	-3%	-3%	-14%	- 8,823
FR	-15%	-6%	-1%	-12%	- 52,635
CZ	0%	-19%	-28%	-6%	- 3,064
PT	-7%	4%	-3%	-6%	- 3,462
PL	7%	0%	3%	-2%	- 3,501
CY	0%	0%	0%	0%	0
MT	0%	0%	0%	0%	0
EE	19%	5%	-3%	2%	164
DE	10%	8%	4%	4%	23,760
SE	2%	3%	1%	5%	6,969
SI	-8%	24%	14%	5%	468
RO	-9%	-3%	-3%	7%	6,439
ES	1%	0%	3%	7%	23,523
UK	2%	0%	0%	8%	34,761
IT	4%	2%	5%	11%	33,371
BE	20%	7%	7%	12%	7,855
FI	17%	16%	19%	18%	13,370
SK	-17%	1%	25%	22%	6,455
NL	3%	5%	11%	23%	26,970
HU	29%	19%	19%	24%	10,429
LU	48%	22%	40%	57%	6,187

B. Additional figures

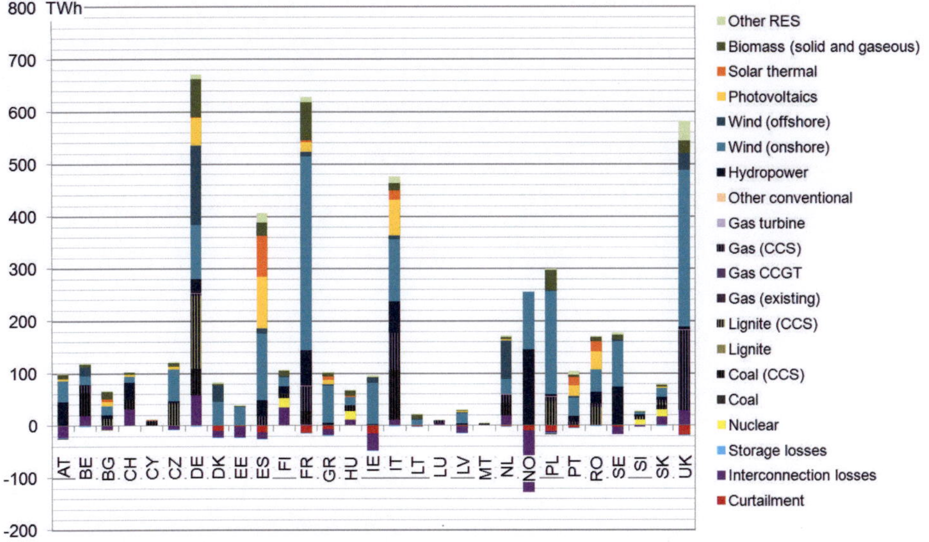

Figure B.1.: Electricity generation, export and import by country, in 2050 of the OPT scenario.

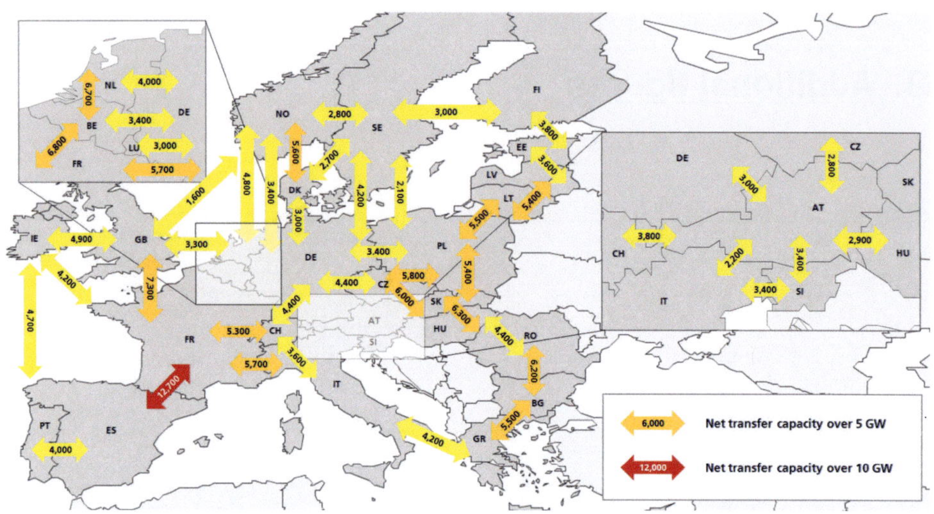

Figure B.2.: NTC of interconnections in 2050 in the OPT scenario (in MW).

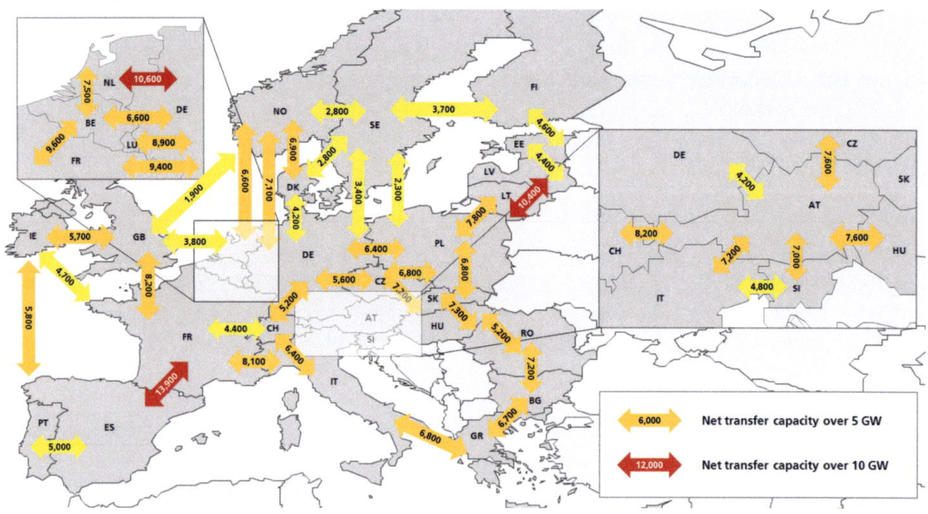

Figure B.3.: NTC of interconnections in 2050 in the NoCCS scenario (in MW).

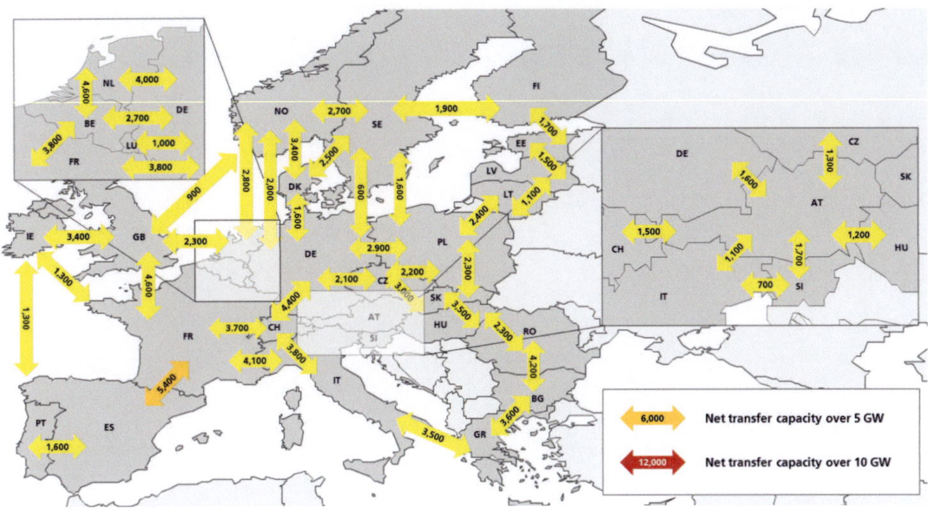

Figure B.4.: NTC of interconnections in 2050 in the GRID scenario (in MW).

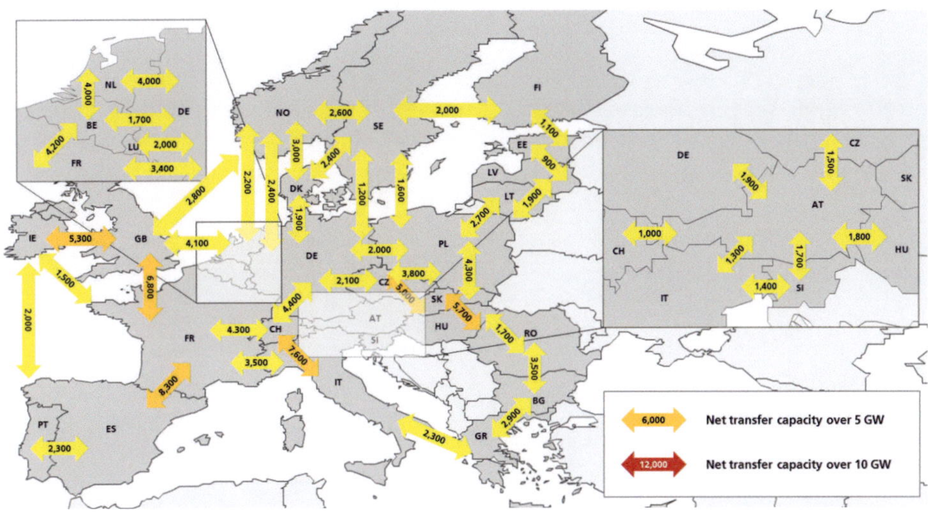

Figure B.5.: NTC of interconnections in 2050 in the EFF scenario (in MW).

Bibliography

Adamec, Marek et al. (2009). "Transformation of Commercial Flows into Physical Flows of Electricity". In: *Acta Polytechnica. 2009, vol. 49.*

Ambec, Stefan and Claude Crampes (2012). "Electricity provision with intermittent sources of energy". In: *Resource and Energy Economics* 34.3, pp. 319–336.

Antoniou, Y. and P. Capros (1999). "Decision support system framework of the PRIMES energy model of the European Commission". In: *International Journal of Global Energy Issues* 12 (1-6), pp. 92-119.

Baíllo, A. (2002). *A methodology to develop optimal schedules and offering strategies for a generation company operating in a short-term electricity market.* Dissertation. Universidad Pontificia Comillas, Madrid.

Balaguer, J. (2011). "Cross-border integration in the European electricity market. Evidence from the pricing behavior of Norwegian and Swiss exporters". In: *Energy Policy* 39.9, pp. 4703–4712.

Benders, J.F. (1962). "Partitioning procedures for solving mixed-variables programming problems". In: *Numerische Mathematik* 4, pp. 238–252.

Beurskens, L.W.M. et al. (2011). *Renewable Energy Projections as Published in the National Renewable Energy Action Plans of the European Member States.* Energy research Centre of the Netherlands. URL: http://www.ecn.nl/docs/library/report/2010/e10069.pdf (visited on 01/18/2013).

Blesl, Markus et al. (2011). "Stromerzeugung 2030 - mit welchen Kosten ist zu rechnen?" In: *Energiewirtschatliche Tagesfragen* (10).

Bode, S. and H. Groscurth (2011). "Elemente für ein zukunftsfähiges Strommarktdesign". In: *Die Zukunft des Strommarktes. Anregungen für den Weg zu 100 Prozent Erneuerbare Energien.* Ed. by D. Schütz and Björn Klusmann. Bundesverband Erneuerbare Energien. Ponte Press, Bochum.

Borenstein, Severin et al. (2000). "The competitive effects of transmission capacity in a deregulated electricity industry". In: *Journal of Economics* 31 (2), pp. 294–325.

Bradley, Stephen P. et al. (1977). *Applied Mathematical Programming*. Addison Wesley, Boston.

Bredin, Don and Cal Muckley (2011). "An emerging equilibrium in the EU emissions trading scheme". In: *Energy Economics* 33.2, pp. 353-362.

Bundesnetzagentur (2012). *"Smart Grid" und "Smart Market". Eckpunktepapier der BundesNetzagentur zu den Aspekten des sich verändernden Energieversorgungssystems*. Report. URL: http://www.bundesnetzagentur.de/ (visited on 01/18/2013).

Bundesverband Geothermie (2010). *Induzierte Seismizität. Position des GtV-BV Geothermie e.V.* Berlin, 7 July 2010. URL: http://www.geothermie.de/uploads/media/GtV-BV_Positionspapier_Seismizitaet_deutsch_0710.pdf (visited on 01/18/2013).

Bunn, Derek W. and Fernando S. Oliveira (2003). "Evaluating Individual Market Power in Electricity Markets via Agent-Based Simulation". In: *Annals of Operations Research* 121 (1). 10.1023/A:1023399017816, pp. 57–77.

Centeno, E. et al. (2007). "Strategic Analysis of Electricity Markets Under Uncertainty: A Conjectured-Price-Response Approach". In: *IEEE Transactions on Power Systems* 22.1, pp. 423–432.

Clímaco, J. and A. T. Almeida (1981). "Multiobjective power systems planning". In: *Proc. of the 3rd International Conference Energy Use Management*. Pergamont Press, Oxford.

Colombo, Marco (2007). *Advances in Iterior Point Methods for Large-Scale Linear Programming*. Dissertation. Univeristy of Edinburgh.

Conejo, A.J. et al. (2002). "Price-taker bidding strategy under price uncertainty". In: *IEEE Transactions on Power Systems* 17.4, pp. 1081–1088.

Connolly, D. et al. (2010). "A review of computer tools for analysing the integration of renewable energy into various energy systems". In: *Applied Energy* 87.4, pp. 1059–1082.

Day, C.J. et al. (2002). "Oligopolistic competition in power networks: a conjectured supply function approach". In: *IEEE Transactions on Power Systems* 17.3, pp. 597–607.

Deckmann, S. et al. (1980). "Numerical Testing of Power System Load Flow Equivalents". In: *IEEE Transactions on Power Apparatus and Systems* 99.6, pp. 2292–2300.

Deetman, Sebastiaan et al. (2012). *Deep greenhouse gas emission reductions: a global bottom-up model approach*. Report for the EU project RESPONSES. PBL - Netherlands Environmental Assessment Agency, Bilthoven, NL.

Destatis (2011). *Monatsbericht über die Elektrizitätsversorgung. Tabelle 11: Leistung und Belastung der Kraftwerke*. not available on http://www.destatis.de/ anymore.

Eberlein, Burkard and Edgar Grande (2005). "Beyond delegation: transnational regulatory regimes and the EU regulatory state". In: *Journal of European Public Policy* 12.1, pp. 89–112.

Ehrenmann, Andreas and Yves Smeers (2011). "Generation Capacity Expansion in a Risky Environment: A Stochastic Equilibrium Analysis". In: *Operations Research* 59.6, pp. 1332–1346.

EirGrid (2010). *System Demand data*. URL: http://www.eirgrid.com/operations/systemperformancedata/systemdemand/.

Elering (2010). *Hourly demand data*. URL: http://elering.ee.

Ellerman, A. Denny and Barbara K. Buchner (2007). "The European Union Emissions Trading Scheme: Origins, Allocation, and Early Results". In: *Review of Environmental Economics and Policy* 1.1, pp. 66–87.

Ellersdorfer, Ingo et al. (2008). *Preisbildungsanalyse des deutschen Elektrizitätsmarktes*. Stuttgart: Institut für Energiewirtschaft und Rationelle Energieanwendung, Universität Stuttgart. URL: http://www.ier.uni-stuttgart.de/publikationen/pb_pdf/Ellersdorfer_Preisbildungsanalyse.pdf (visited on 01/18/2013).

EMCAS (2008). *Electricity Market Complex Adaptive System (EMCAS). Model introduction*. URL: http://www.dis.anl.gov/projects/emcas.htm (visited on 01/18/2013).

Energy and Ecology Blog (2012). *Levelized Costs of Electricity Generation (LCOE) - 2012 Update.* URL: http://energy-ecology.blogspot.de/2012/03/levelized-costs-of-electricity.html (visited on 01/18/2013).

ENTSO-E (2012a). *ENTSO-E Transparency Platform.* URL: http://http://www.entsoe.net (visited on 01/18/2013).

ENTSO-E (2012b). *Online documentation of historic NTC Matrixes.* URL: https://www.entsoe.eu/resources/ntc-values/ntc-matrix/ (visited on 01/18/2013).

Enzensberger, N. (2003). "Entwicklung und Anwendung eines Strom- und Zertifikatemarktmodells für den europäischen Stromsektor". Universität Karlsruhe.

Erev, Ido and Alvin E. Roth (1998). "Predicting How People Play Games: Reinforcement Learning in Experimental Games with Unique, Mixed Strategy Equilibria". In: *The American Economic Review* 88, pp. 848–881.

Euracoal (2011). *Country Profiles.* URL: http://www.euracoal.be.

Eurelectric (2009). *Power Choices - Pathways to Carbon-Neutral Electricity in Europe by 2050.* URL: www.eurelectric.org/PowerChoices2050 (visited on 01/18/2013).

ECF – European Climate Foundation (2010). *Roadmap 2050. A practical guide to a prosperous, low-carbon Europe.* Vol. Volume 1 -. URL: http://www.roadmap2050.eu/Volume1_fullreport_PressPack.pdf (visited on 01/18/2013).

European Commission (1996a). *Directive 96/92/EC of the European Parliament and of the Council of 19 December 1996 concerning common rules for the internal market in electricity.*

European Commission (1996b). *Energy for the Future: Renewable Sources of Energy, Green Paper for a Community Strategy. COM (96) 576 final, 20 November 1996.*

European Commission (1997). *Energy for the future: Renewable sources of energy. White Paper for a Community strategy and action plan. COM(97) 599 final.*

European Commission (2001). *Directive 2001/77/EC of the European Parliament and of the Council of 27 September 2001 on the promotion of electricity produced from renewable energy sources in the internal electricity market.*

European Commission (2003). *Directive 2003/87/EC f the European Parliament and of the Council of 13 October 2003 establishing a scheme for greenhouse gas emission allowance trading within the Community and amending Council Directive 96/61/EC.*

European Commission (2007). *DG Competition report on energy sector inquiry, SEC (2006)1724.* Final report.

European Commission (2009a). *Directive 2009/28/EC of the European Parliament and of the Council of 23 April 2009 on the promotion of the use of energy from renewable sources and amending and subsequently repealing Directives 2001/77/EC and 2003/30/EC.*

European Commission (2009b). *Directive 2009/29/EC f the European Parliament and of the Council of 23 April 2009 amending Directive 2003/87/EC so as to improve and extend the greenhouse gas emission allowance trading scheme of the Community.*

European Commission (2009c). *Directive 2009/31/EC of 23 April 2009 on the geological storage of carbon dioxide and amending Council Directive 85/337/EEC, European Parliament and Council Directives 2000/60/EC, 2001/80/EC, 2004/35/EC, 2006/12/EC, 2008/1/EC and Regulation (EC) No 1013/2006 2003/54/EC.*

European Commission (2009d). *Directive 2009/72/EC of the European Parliament and of the Council of 13 July 2009 concerning common rules for the internal market in electricity and repealing Directive 2003/54/EC.*

European Commission (2009e). *Regulation (EC) No 713/2009 of the European Parliament and of the Council of 13 July 2009 establishing an Agency for the Cooperation of Energy Regulators.*

European Commission (2009f). *Regulation (EC) No 714/2009 of the European Parliament and of the Council of 13 July 2009 on conditions for access to the network for cross-border exchanges in electricity and repealing Regulation (EC) No 1228/2003.*

European Commission (2011a). *Communication from the Commission to the European Parliament, the Council, the Economic and Social Committee and the Committee of the Regions. A Roadmap for moving to a competitive low carbon economy in 2050. COM(2011) 112 final.*

European Commission (2011b). *Communication from the Commission to the European Parliament, the Council, the Economic and Social Committee and the Com-*

mittee of the Regions of 8 March 2011. Energy Efficiency Plan 2011. COM(2011) 109 final.

European Commission (2011c). *Communication from the Commission to the European Parliament, the Council, the European Economic and Social Committee and the Committee of the Regions. Energy Roadmap 2050. COM/2011/0885 final.*

European Commission (2011d). *Impact Assessment accompanying the document Energy Roadmap 2050. SEC(2011) 1565/2 Part 1/2.* Commission staff working paper.

European Commission (2011e). *National renewable energy action plans.* URL: http://ec.europa.eu/energy/renewables/action_plan_en.htm (visited on 01/18/2013).

ENTSO-E – European Network of Transmission System Operators for Electricity (2001). *Definitions of Transfer Capacities in liberalised Electricity Markets.* Final Report. Not vailable anymore on the original website, but a copy can be found on: http://www.docstoc.com/docs/44746116/Definitions-of-Transfer-Capacities-in-liberalised-Electricity-Markets.

ENTSO-E – European Network of Transmission System Operators for Electricity (2012c). *10-Year Network Development Plan 2012.* URL: https://www.entsoe.eu/major-projects/ten-year-network-development-plan/ (visited on 01/18/2013).

European Union (2007). *Treaty of Lisbon amending the Treaty on European Union and the Treaty establishing the European Community, signed at Lisbon, 13 December 2007. Nocie No. 2007/C 3006/01.* Official Journal of the European Union.

Eurostat (2012). *Database on environment and energy.* URL: http://epp.eurostat.ec.europa.eu/portal/page/portal/statistics/search_database (visited on 01/18/2013).

Ferreira Dias, Marta (2011). *Integration of European electricity markets.* Dissertation, University of Warwick.

Fichtner, W. (1999). *Strategische Optionen der Energieversorger zur CO_2-Minderung: ein Energie- und Stoffflussmodell zur Entscheidungsunterstützung.* Erich Schmidt Verlag, Berlin.

Fink, Bernhard (2011). "Does large turbine technically feasible and more cost efficient: The Enercon E-126 - From wind turbine to power plant". In: *Proceedins of the "Global Renewable Energy Development 2011".*

Finkenrath, Matthias (2011). "Cost and performance of carbon dioxide capture from power generation". In: *IEA Working Paper*. URL: www.iea.org/publications/freepublications/publication/name,3950,en.html (visited on 01/18/2013).

Fischedick, Manfred et al. (2012). *Power sector decarbonisation: Metastudy*. Final study for the SEFEP funded project 11-01. Öko-Institut e.V. URL: www.oeko.de/oekodoc/1352/2012-004-en.pdf (visited on 01/18/2013).

Fleiter, Tobias et al. (2009). "Costs and potentials of energy savings in European industry - a critical assessment of the concept of conservation supply curves". In: *Proceedings of the ECEEE 2009 Summer study*.

Fleten, S.-E. et al. (2007). "Optimal investment strategies in decentralized renewable power generation under uncertainty". In: *Energy* 32.5, pp. 803–815.

Fleten, Stein-Erik et al. (2002). "Hedging electricity portfolios via stochastic programming". In: *Decision Making under Uncertainty Energy and Power*. Vol. 128. IMA Vol. Math. Appl. Springer, pp. 71–93.

Forrester, J. W. (1958). "Industrial Dynamics: A Major Breakthrough for Decision Makers". In: *Harvard Business Review* 36.4, pp. 37–66.

France, N., ed. (2000). *Special Issue: On the Rebound: The Interaction of Energy Efficiency, Energy Use and Economic Activity*. Vol. 28, Numbers 6-7.

Franke, Günter and Herbert Hax (2009). *Finanzwirtschaft des Unternehmens und Kapitalmarkt*. 7th ed. Springer Berlin Heidelberg.

Fuss, Sabine and Jana Szolgayová (2010). "Fuel price and technological uncertainty in a real options model for electricity planning". In: *Applied Energy* 87.9, pp. 2938–2944.

Fuss, Sabine, Jana Szolgayová, et al. (2012). "Renewables and climate change mitigation: Irreversible energy investment under uncertainty and portfolio effects". In: *Energy Policy* 40, pp. 59–68.

Genoese, Fabio, Massimo Genoese, et al. (2012). "Medium-term Flexibility Options in a Power Plant Portfolio - Energy Storage Units vs. Thermal Units". In: *Proceedings of the 9th International Conference on the European Energy Market*.

Genoese, Fabio and Martin Wietschel (2011). "Großtechnische Stromspeicheroptionen im Vergleich". In: *Energiewirtschaftliche Tagesfragen* 61.6, pp. 26–31.

Giebel, G. (2005). *Wind power has a capacity credit: A catalogue of 50+ supporting studies*. RisøNational Laboratory. URL: ejournal.windeng.net/3/ (visited on 01/18/2013).

Greenpeace (2012). *The EPR nuclear reactor. A dangerous waste of time and money*. Briefing. URL: http://www.greenpeace.org/international/en/publications/ Campaign-reports/Nuclear-reports/EPR-Nuclear-Reactor-2012/ (visited on 01/18/2013).

Gross, G. and D. J. Finlay (1996). "Optimal bidding strategies in competitive electricity markets". In: *Proceedings of the 12th PSCC Conference, Germany*.

Haas, R. et al. (2004). "How to promote renewable energy systems successfully and effectively". In: *Energy Policy* 32.6, pp. 833–839.

Hamelinck, Carlo et al. (2012). *Renewable energy progress and biofuels sustainability*. Report for the European Commission.

Hasani-Marzooni, M. and S. H. Hosseini (2011). "Dynamic model for market-based capacity investment decision considering stochastic characteristic of wind power". In: *Renewable Energy* 36.8, pp. 2205–2219.

Hearps, Patrick and Dylan McConnell (2011). *Renewable Energy Technology Cost Review*. Technical Paper Series. Melbourne Energy Institute.

Heide, Dominik et al. (2010). "Seasonal optimal mix of wind and solar power in a future, highly renewable Europe". English. In: *Renewable Energy* 35.11, pp. 2483-2489.

Held, Anne Mirjam (2011). *Modelling the future development of renewable energy technologies in the European electricity sector using agent-based simulation*. Stuttgart: Dissertation. Fraunhofer Verlag, Stuttgart.

Herring, Horace (2006). "Energy efficiency - a critical view". In: *Energy* 31.1, pp. 10–20.

Heuck, Klaus et al. (2007). *Elektrische Energieversorgung: Erzeugung, Übertragung und Verteilung elektrischer Energie für Studium und Praxis*. 7th ed. Vieweg+Teubner Verlag.

Hobbs, B.F. et al. (2000). "Strategic gaming analysis for electric power systems: an MPEC approach". In: *IEEE Transactions on Power Systems* 15.2, pp. 638–645.

Howarth, Richard B. (2006). "Against High Discount Rates". In: *Advances in the Economics of Environmental Resources* 5, pp. 99–120.

Huisman, Ronald and Mehtap Kiliç (in press). "A history of European electricity day-ahead prices". In: *Applied Economics* 45.18, pp. 2683–2693.

IEA Wind Task 25 (2009). *Design and operation of power systems with large amounts of wind power,* Final report. Finland, VTT Research Notes 2493, VTT [Online]. URL: http://www.vtt.fi/inf/pdf/tiedotteet/2009/T2493.pdf (visited on 01/18/2013).

energynautics and EWI – Institute of Energy Economics at the University of Cologne (2011). *Roadmap 2050 - a closer look. Cost-efficient RES-E penetration and the role of grid extensions.* Final report. URL: http://www.ewi.uni-koeln.de/fileadmin/user_upload/Publikationen/Studien/Politik_und_Gesellschaft/2011/Roadmap_2050_komplett_Endbericht_Web.pdf (visited on 01/18/2013).

IPCC – Intergovernmental Panel on Climate Change (2007a). *Climate Change 2007 - Mitigation of Climate Change: Working Group III contribution to the Fourth Assessment Report of the IPCC.* 1st ed. Cambridge University Press.

IPCC – Intergovernmental Panel on Climate Change (2007b). *Climate Change 2007 - The Physical Science Basis: Working Group I Contribution to the Fourth Assessment Report of the IPCC.* 1 Pap/Cdr. Cambridge University Press.

IPCC – Intergovernmental Panel on Climate Change (2008). *Climate Change 2007 - Impacts, Adaptation and Vulnerability: Working Group II contribution to the Fourth Assessment Report of the IPCC.* 1 Pap/Cdr. Cambridge University Press.

IEA – International Energy Agency (2010). *Projected Costs of Generating Electricity - 2010 Edition.* Paris, France.

IEA – International Energy Agency (2011a). *CO_2 Emissions from Fuel Combustion 2011.* 2011 ed. Paris, France.

IEA – International Energy Agency (2011b). *World Energy Outlook 2011.* 1st ed. Paris, France.

IRENA – International Renewable Energy Agency (2012a). *Renewable Energy Cost Analysis - Concentrating Solar Power.* IRENA Working Paper. URL: http://www.irena.org/DocumentDownloads/Publications/RE_Technologies_Cost_Analysis-CSP.pdf.

IRENA – International Renewable Energy Agency (2012b). *Renewable Energy Cost Analysis - Solar Photovoltaics*. IRENA Working Paper. URL: http://www.irena. org / DocumentDownloads / Publications / RE_ Technologies_ Cost_ Analysis - SOLAR_PV.pdf (visited on 01/18/2013).

IRENA – International Renewable Energy Agency (2012c). *Renewable Energy Cost Analysis - Wind Power*. IRENA Working Paper. URL: http://www.irena.org/ DocumentDownloads/Publications/RE_Technologies_Cost_Analysis-WIND_ POWER.pdf (visited on 01/18/2013).

IPCC (2012). *Renewable Energy Sources and Climate Change Mitigation. Special Report of the Intergovernmental Panel on Climate Change.* Ed. by O. Edenhofer et al. New York, USA: Cambridge University Press.

Jacobsson, Staffan and Volkmar Lauber (2006). "The politics and policy of energy system transformation - explaining the German diffusion of renewable energy technology". In: *Energy Policy* 34.3, pp. 256–276.

Jäger-Waldau, Arnulf et al. (2011). "Renewable electricity in Europe". In: *Renewable and Sustainable Energy Reviews* 15.8, pp. 3703–3716.

Jamasb, T and M. G. Pollitt (2005). "Electricity market reform in the European Union: Review of progress toward liberalization & integration". In: *Working Paper*. Massachusetts Institute of Technology, Center for Energy and Environmental Policy Research, pp. 11-41.

Jin, Shan and S.M. Ryan (2011). "Capacity Expansion in the Integrated Supply Network for an Electricity Market". In: *IEEE Transactions on Power Systems* 26.4, pp. 2275–2284.

Jones, Christopher and William Webster (2006). *EU Energy Law, Volume I: The Internal Energy Market: 1*. Updated. Claeys & Casteels.

Kagiannas, Argyris G. et al. (2004). "Power generation planning: a survey from monopoly to competition". In: *International Journal of Electrical Power & Energy Systems* 26.6, pp. 413–421.

Kahn, Edward P. (1998). "Numerical Techniques for Analyzing Market Power in Electricity". In: *The Electricity Journal* 11.6, pp. 34–43.

Karan, Mehmet Baha and Hasan Kazdağli (2011). "The Development of Energy Markets in Europe". In: *Financial Aspects in Energy: A European Perspective*. Ed. by André Dorsman et al. 1st ed. Springer.

Kazempour, S.J. and A.J. Conejo (2012). "Strategic Generation Investment Under Uncertainty Via Benders Decomposition". In: *IEEE Transactions on Power Systems* 27.1, pp. 424–432.

Kesicki, Fabian and Gabrial Anandarajah (2011). "The role of energy-service demand reduction in global climate change mitigation: Combining energy modelling and decomposition analysis". In: *Energy Policy* 39.11, pp. 7224–7233.

Kettunen, J. et al. (2010). "Optimization of Electricity Retailer's Contract Portfolio Subject to Risk Preferences". In: *IEEE Transactions on Power Systems* 25.1, pp. 117–128.

KfW/ZEW (2012). *CO₂ Barometer 2012*. URL: http://www.zew.de/en/publikationen/ CO2panel.php (visited on 11/04/2012).

Kirkels, Arjan F. (2012). "Discursive shifts in energy from biomass: A 30 year European overview". In: *Renewable and Sustainable Energy Reviews* 16.6, pp. 4105–4115.

Klein, Arne et al. (2010). *Evaluation of different feed-in tariff design options. Best practice paper for the International Feed-In Cooperation. 3rd edition.* URL: http://www.feed-in-cooperation.org/wDefault_7/download-files/research/Best_practice_Paper_3rd_edition.pdf (visited on 01/18/2013).

Koch, M. et al. (2001). *A Systematic Analysis of the Characteristics of Energy Models with Regard to their Suitability for Practical Policy Recommendations for Developing Future Strategies to Mitigate Climate Change.* Final report. Berlin: German Federal Environmental Agency (Umweltbundesamt).

Koopman, Siem Jan et al. (2007). "Periodic seasonal Reg-ARFIMA-GARCH models for daily electricity spot prices". English. In: *Journal of the American Statistical Association* 102, pp. 16-27.

Korpås, M. et al. (2007). *TradeWind. Grid modelling and power system data.* Deliverable. URL: http://www.trade-wind.eu/ (visited on 10/20/2011).

Krewitt, W. et al. (2009). *Role and Potential of Renewable Energy and Energy Efficiency for Global Energy Supply.* Climate Change 18/2009, Federal Environment Agency, Dessau-Roßlau, Germany.

Krey, Volker (2006). "Vergleich kurz- und langfristig ausgerichteter Optimierungsansätze mit einem multi-regionalen Energiesystemmodell unter Berücksichtigung stochastischer Parameter". Dissertation, Ruhr-Universität Bochum.

Kumbaroğlu, Gürkan et al. (2008). "A real options evaluation model for the diffusion prospects of new renewable power generation technologies". In: *Energy Economics* 30.4, pp. 1882–1908.

Leigh and Tesfatsion (2006). "Chapter 16 Agent-Based Computational Economics: A Constructive Approach to Economic Theory". In: *Handbook of Computational Economics. Agent-based computational economics*. Ed. by L. Tesfatsion and K.L. Judd. Vol. 2. Handbook of Computational Economics. Elsevier, pp. 831–880.

Leontief, W. (1966). *Input-Output Economics*. Oxford University Press.

Lienert, Martin and Stefan Lochner (2012). "The importance of market interdependencies in modeling energy systems - The case of the European electricity generation market". In: *International Journal of Electrical Power & Energy Systems* 34.1, pp. 99–113.

Ludig, Sylvie et al. (2011). "Fluctuating renewables in a long-term climate change mitigation strategy". In: *Energy* 36.11, pp. 6674–6685.

DLR – Luft- und Raumfahrt (German Aerospace Center), Deutsches Zentrum für (2006). *TRANS-CSP. Trans-Mediterranean interconnection for Concentrating Solar Power*. URL: www.dlr.de/tt/desktopdefault.aspx/tabid-2885/4422_read-6588 (visited on 01/18/2013).

Lund, Henrik et al. (2007). "Two energy system analysis models: A comparison of methodologies and results". In: *Energy* 32.6, pp. 948–954.

McKinsey & Company (2010). *Transformation of Europes power system until 2050, including specific considerations for Germany*.

Meadows, Donella H. et al. (1972). *Limits to Growth*. 1st ed. Universe Books, New York.

Meeus, Leonardo (2011). "Why (and how) to regulate power exchanges in the EU market integration context?" In: *Energy Policy* 39.3, pp. 1470–1475.

Meinshausen, Malte (2005). "On the Risk of Overshooting 2°C." In: *Scientific Symposium "Avoiding Dangerous Climate Change", Exeter, 1-3 February 2005*.

Messner, S. et al. (1996). "A stochastic version of the dynamic linear programming model MESSAGE III". In: *Energy* 21.9, pp. 775–784.

Meza, J. L. C. et al. (2007). "A Model for the Multiperiod Multiobjective Power Generation Expansion Problem". In: *IEEE Transactions on Power Systems* 22.2, pp. 871–878.

Mitra-Kahn, B. H. (2008). "Debunking the Myths of Computable General Equilibrium Models". *Schwartz Center for Economic Policy Analysis Working Paper*. New York.

Möst, Dominik and Dogan Keles (2010). "A survey of stochastic modelling approaches for liberalised electricity markets". In: *European Journal of Operational Research* 207.2, pp. 543–556.

Nathani, Carsten (2008). *Modellierung des Strukturwandels beim Üergang zu einer materialeffizienten Kreislaufwirtschaft: Kopplung eines Input-Output-Modells mit einem Stoffstrommodell am Beispiel der Wertschöpfungskette "Papier"*. Physica, Berlin.

National Grid (2010). *Metered half-hourly electricity demands*. URL: http://www.nationalgrid.com/uk/Electricity/Data/Demand+Data/ (visited on 01/18/2013).

Newell, Richard G. and William A. Pizer (2004). "Uncertain discount rates in climate policy analysis". In: *Energy Policy* 32.4, pp. 519–529.

Nguyen, D.H.M. and K.P. Wong (2002). "Natural dynamic equilibrium and multiple equilibria of competitive power markets". In: *IEE Proceedings - Generation, Transmission and Distribution* 149.2, pp. 133–138.

Nicolosi, M. (2012). "The Economics of Renewable Electricity Market Integration. An Empirical and Model-Based Analysis of Regulatory Frameworks and their Impacts on the Power Market". Dissertation, University of Cologne.

Nishimura, F. et al. (1993). "Benefit optimization of centralized and decentralized power systems in a multi-utility environment". In: *IEEE Transactions on Power Systems* 8.3, pp. 1180–1186.

Nuclear Engineering International (2011). *EDF delays Flamanville 3 EPR project*. 20 July 2011. URL: http://www.neimagazine.com/story.asp?sectioncode=132&storyCode=2060192 (visited on 01/18/2013).

Nyober (1997). "Simulating evolution of technology: An aid to energy policy analysis." School of Resource and Environmental Management, Simon Fraser University, Vancouver.

Oda, Junichiro and Keigo Akimoto (2011). "An analysis of CCS investment under uncertainty". In: *Energy Procedia* 4, pp. 1997–2004.

Oettinger, G. (2011). *Press conference by Günther Oettinger, European Commissioner for Energy, for the publication of the "Energy Roadmap 2050".* URL: http://ec.europa.eu/avservices/player/streaming.cfm?type=ebsvod\&sid=193030 (visited on 05/03/2012).

Olmos, L. et al. (2011). "Barriers to the implementation of response options aimed at mitigating the impact of wind power on electricity systems". In: *Wind Energy* 14.6, pp. 781-795.

Olsina, Fernando et al. (2006). "Modeling long-term dynamics of electricity markets". In: *Energy Policy* 34.12, pp. 1411–1433.

Panzer, Christian (2012). "Investment costs of renewable energy technologies under consideration of volatile raw material prices," Dissertation. Technical Unicersity Vienna.

Pfluger, B. (2009). "Agentenbasierte Simulation der Auswirkungen eines Imports von Strom aus erneuerbaren Energiequellen in Nordafrika in den italienischen Markt". In: *6. Internationale Energiewirtschaftstagung an der TU Wien, Vienna.*

Pielow, Johann-Christian and Britta Janina Lewendel (2011). "The EU Energy Policy After the Lisbon Treaty". In: *Financial Aspects in Energy.* Ed. by André Dorsman et al. Springer Berlin Heidelberg, pp. 147–165. (Visited on 09/21/2012).

Platts (2010). *The UDI World Electric Power Plants Database.* URL: http://www.platts.com/Products/worldelectricpowerplantsdatabase (visited on 01/18/2013).

Plazas, M.A. et al. (2005). "Multimarket optimal bidding for a power producer". In: *IEEE Transactions on Power Systems* 20.4, pp. 2041–2050.

Pollitt, M. G. (2009). "Electricity Liberalisation in the European Union: A Progress Report". *Cambridge Working Papers in Economics.* URL: http://ideas.repec.org/p/cam/camdae/0953.html (visited on 09/06/2012).

Pöyry (2011). *The challenges of intermittency in North West European power markets. The impacts when wind and solar deployment reach their target levels. Ex-*ctract from a study. URL: http://www.poyry.com/sites/default/files/143.pdf (visited on 02/21/2012).

PwC – PricewaterhouseCoopers, Potsdam-Institut für Klimaforschung, International Institute for Applied Science Analysis, European Climate Forum (2010). *100% reeneable electricty. A roadmap to 2050 for Europe and North Africa.* London, Potsdam, Laxenburg. URL: http://www.pwc.co.uk/electricity-gas/publications/100-percent-renewable-electricity.jhtml (visited on 01/18/2013).

Pront-van Bommel, Simone (2011). "The Development of the European Electricity Market in a Juridical No Mans Land". In: *Financial Aspects in Energy.* Ed. by André Dorsman et al. Springer Berlin Heidelberg, pp. 167–193.

Purchala, K., E. Haesen, et al. (2005). "Zonal network model of European interconnected electricity network". In: *International CIGRE/IEEE Symposium 2005,* pp. 362–369.

Purchala, K., L. Meeus, et al. (2005). "Usefulness of DC power flow for active power flow analysis". In: *Power Engineering Society General Meeting, 2005. IEEE,* 454–459 Vol. 1.

Ragwitz, M. et al. (2011). *Assessment of National Renewable Energy Action Plans (interim status).* URL: http://www.eufores.org/fileadmin/eufores/Projects/REPAP_2020/Assessment_of_NREAPs__REPAP_report_-_interim_status_.pdf (visited on 01/18/2013).

Ragwitz et al. (2012). *RE-Shaping: Shaping an effective and efficient European renewable energy market.* Final Report. URL: www.reshaping-res-policy.eu/ (visited on 01/18/2013).

Rajamaran, R. et al. (2001). "Optimal self-commitment under uncertain energy and reserve prices". In: *The Next Generation of Unit Commitment Models.* Kluwer Acedemic Publishers, Boston.

Rhodes, James and David Keith (2008). "Biomass with capture: negative emissions within social and environmental constraints: an editorial comment". In: *Climatic Change* 87 (3), pp. 321–328.

Rogge, Karoline S. et al. (2011). "The innovation impact of the EU Emission Trading System - Findings of company case studies in the German power sector". In: *Ecological Economics* 70.3, pp. 513-523.

Rosen, Johannes (2008). *The future role of renewable energy sources in European electricity supply: A model-based analysis for the EU-15.* Dissertation, Universität Karlsruhe.

Sauer, P.W. et al. "Extended factors for linear contingency analysis". In: *Proceedings of the 34th Annual Hawaii International System Sciences.*

Schaber, Katrin et al. (2012). "Transmission grid extensions for the integration of variable renewable energies in Europe: Who benefits where?" In: *Energy Policy* 43, pp. 123–135.

Schiavone, F. (2010). "The liberalisation and diversification of the European electricity industry: A country-based view". In: *Journal of General Management* 36.3, pp. 63–79.

Schliesinger, M. et al. (2010). *Energieszenarien für ein Energiekonzept der Bundesregierung.* Report for the German Federal Ministry of Economics and Technology. URL: http://www.bmu.de/fileadmin/bmu-import/files/pdfs/allgemein/application/pdf/energieszenarien_2010.pdf (visited on 01/18/2013).

Schubert, Gerda (2012). "Modeling hourly electricity generation from PV and wind plants in Europe". In: *Proceedings of the 9th International Conference on the European Energy Market.*

Sensfuß, Frank (2007). "Assessment of the impact of renewable electricity generation on the German electricity sector: An agent-based simulation approach". Dissertation. VDI Verlag, Düsseldorf.

Severance, Craig A. (2009). "Business Risks and Costs of New Nuclear Power". In: URL: http://thinkprogress.org/wp-content/uploads/2009/01/nuclear-costs-2009.pdf (visited on 02/13/2009).

Sioshansi, Fereidoon P. (2006). "Electricity Market Reform: What Have We Learned? What Have We Gained?" In: *The Electricity Journal* 19.9, pp. 70–83.

Steinhilber, Simone et al. (2011). *Indicators assessing the performance of renewable energy support policies in 27 Member States.* Project report. Fraunhofer ISI, Ecofys. URL: www.reshaping-res-policy.eu/ (visited on 01/18/2013).

Sterman, J. (2000). *Business dynamics: Sytems thinking and modeling for a complex world.* Irwin/McGraw-Hill, Boston.

Stern, Nicholas (2007). *The Economics of Climate Change: The Stern Review.* Cambridge University Press.

Sun, Junjie and Leigh Tesfatsion (2007). "Dynamic Testing of Wholesale Power Market Designs: An Open-Source Agent-Based Framework". In: *Computational Economics* 30.3, pp. 291–327.

Targosz, R. (2008). *Network Losses.* Webinar April 11, 2008. URL: http://www.scribd.com/doc/2585592/Network-Losses (visited on 01/18/2013).

ter Keurst, M. (2011). "The Effect of European Liberalisation Policy on Electricity Prices. An Empirical Study". Master thesis. Tilburg University.

Teske, Sven, Josche Muth, et al. (2012). *energy [r]evolution. A Sustainable World Energy Outlook. 4th edition 2012 World Energy Scenario.* URL: www.greenpeace.org/international/en/publications/Campaign-reports/Climate-Reports/Energy-Revolution-2012/ (visited on 01/18/2013).

Teske, Sven, Arthouros Zervos, et al. (2010). *energy [r]evolution. A Sustainable World Energy Outlook. 3rd edition.* Greenpeace International and European Renewable Energy Council. URL: http://www.greenpeace.org/international/en/publications/reports/Energy-Revolution-A-Sustainable-World-Energy-Outlook/ (visited on 01/18/2013).

Teske, Sven, A. Zervos, et al. (2007). *energy [r]evolution. A Sustainable World Energy Outlook.* Greenpeace and EREC.

Tilman, David et al. (2009). "Beneficial Biofuels - The Food, Energy, and Environment Trilemma". In: *Science* 325.5938, pp. 270–271.

Torre, S. de la et al. (2008). "Transmission Expansion Planning in Electricity Markets". In: *IEEE Transactions on Power Systems* 23.1, pp. 238–248.

Tóth, F.L. (2000). "Intergenerational equity and discounting". In: *Integrated Assessment* 1, pp. 127–136.

TRNSYS (2011). *TRNSYS17. Model documentation website of the University of Wisconsin, Madison.* URL: http://sel.me.wisc.edu/trnsys/features/features.html (visited on 01/18/2013).

UNFCC (2011). *Greenhouse Gas Inventory Data - Detailed data by Party.* URL: http://unfccc.int/di/DetailedByParty.do (visited on 01/18/2013).

Unger, G. (2002). "Hedging Strategy and Electricity Contract Engineering". Dissertation, Swiss Federal Institute of Technology, Zurich.

United Nations (2009). *Report of the Conference of the Parties on its fifteenth session, held in Copenhagen from 7 to 19 December 2009.*

United Nations (2010). *Compilation of pledges for emission reductions and related assumptions provided by Parties to date and the associated emission reductions. Updated July 2010.*

Ventosa, M. et al. (2005). "Electricity market modeling trends". In: *Energy Policy* 33.7, pp. 897–913.

Verbund AG (2009). *Info am Draht. Starker Strom für unser Land. Ein Überblick über den Bau der 380-kV-Steiermarkleitung.* URL: http://portal.tugraz.at/portal/page/portal/Files/i4340/daten/apg_steiermarkleitung.pdf (visited on 10/20/2012).

Vogstad, Klaus-Ole (2004). "A system dynamics analysis of the Nordic electricity market: The transition from fossil fuelled toward a renewable supply within a liberalised electricity market". Dissertation, Norwegian University of Science and Technology, Trondheim.

von Hirschhausen, Christian et al. (2012). "How a "Low Carbon" Innovation Can Fail–Tales from a "Lost Decade" for Carbon Capture, Transport, and Sequestration (CCTS)". In: *Economics of Energy & Environmental Policy* 1.2.

Weber, Christoph and Jan Swider (2004). "Power plant investments under fuel and carbon price uncertainty". In: *6th IAEE European Conference on Modelling in Energy Economics and Policy.*

Weidlich, Anke and Daniel Veit (2008). "A critical survey of agent-based wholesale electricity market models". In: *Energy Economics* 30.4, pp. 1728–1759.

Wen, Fushuan and A.K. David (2001). "Optimal bidding strategies and modeling of imperfect information among competitive generators". In: *IEEE Transactions on Power Systems* 16.1, pp. 15–21.

Wesselink, Bart et al. (2010). *Energy Savings 2020. How to triple the impact of energy saving policies in Europe.* Final report. URL: www.roadmap2050.eu/contributing_studies (visited on 01/18/2013).

Widén, J. (2011). "Correlations Between Large-Scale Solar and Wind Power in a Future Scenario for Sweden". In: *IEEE Transactions on Sustainable Energy* 2.2, pp. 177–184.

Wiesenthal, Tobias et al., eds. (2006). *How much bioenergy can Europe produce without harming the environment?* EEA Report No 7/2006.

Wietschel, Martin (1995). "Die Wirtschaftlichkeit klimaverträglicher Energieversorgung. Entwicklung und Bewertung von CO_2-Minderungsstrategien in der Energieversorgung und -nachfrage". University of Karlsruhe, Institute for Industrial Production.

Winkler, J. (2012). "Electricity market design for 100% renewable electricity in Germany". In: *12. Symposium Energieinnovation, Graz.*

Woolridge, M.J. (2005). *An Introduction to MultiAgent Systems.* John Wiley & Sons, Chichester.

Wright, Margaret H. (2004). "The interior-point revolution in optimization: History, recent developments, and lasting consequences". In: *Bulletin of the American Mathematical Society* 42.01, pp. 39–56.

Wuppertal Institut – Wuppertal Institut für Klima, Umwelt, Energie) (2010). *Renewable Energies (RE) in Comparison to Carbon Capture and Storage (CCS): An Update.* URL: http://wupperinst.org/en/projects/project-details/wi/p/s/pd/251/ (visited on 01/18/2013).

Yang, Ming et al. (2008). "Evaluating the power investment options with uncertainty in climate policy". In: *Energy Economics* 30.4, pp. 1933–1950.

Zervos, A. et al. (2010). *RE-Thinking 2050 - A 100% Renewable Energy Vision for the European Union.* European Renewable Energy Council. URL: http://www.rethinking2050.eu/fileadmin/documents/ReThinking2050_full_version_final.pdf (visited on 01/18/2013).

Zickfeld, Florian and Aglaia Wieland (2012). *EUMENA 2050 - Powered by renewable energy.* Project report of the Desertec industrial initiative (Dii). URL: http://www.dii-eumena.com/fileadmin/flippingbooks/dp2050_pb_engl_web.pdf (visited on 01/18/2013).